10
Topics in Organometallic Chemistry

Editorial Board:
J. M. Brown · P. H. Dixneuf · A. Fürstner
L. S. Hegedus · P. Hofmann · P. Knochel
S. Murai · M. Reetz · G. van Koten

Topics in Organometallic Chemistry

Recently Published and Forthcoming Volumes

Metal Carbenes in Organic Synthesis
Volume Editor: K. H. Dötz
Vol. 13, 2004

Theoretical Aspects of Transition Metal Catalysis
Volume Editor: G. Frenking
Vol. 12, 2005

Ruthenium Catalysts and Fine Chemistry
Volume Editors:
C. Bruneau, P. H. Dixneuf
Vol. 11, 2004

New Aspects of Zirconium Containing Organic Compounds
Volume Editor: I. Marek
Vol. 10, 2005

CVD Precursors
Volume Editor: R. Fischer
Vol. 9, 2005

Metallocenes in Regio- and Stereoselective Synthesis
Volume Editor: T. Takahashi
Vol. 8, 2005

Transition Metal Arene π-Complexes in Organic Synthesis and Catalysis
Volume Editor: E. P. Kündig
Vol. 7, 2004

Organometallics in Process Chemistry
Volume Editor: R. D. Larsen
Vol. 6, 2004

Organolithiums in Enantioselective Synthesis
Volume Editor: D. M. Hodgson
Vol. 5, 2003

Organometallic Bonding and Reactivity: Fundamental Studies
Volume Editors: J. M. Brown, P. Hofmann
Vol. 4, 1999

Activation of Unreactive Bonds and Organic Synthesis
Volume Editor: S. Murai
Vol. 3, 1999

Lanthanides: Chemistry and Use in Organic Synthesis
Volume Editor: S. Kobayashi
Vol. 2, 1999

Alkene Metathesis in Organic Synthesis
Volume Editor: A. Fürstner
Vol. 1, 1998

New Aspects of Zirconium Containing Organic Compounds

Volume Editor: I. Marek

With contributions by
P. Bertus · N. Chinkov · S. A. Cummings · M. S. Eisen · A. Lisovskii ·
I. Marek · M. Mori · J. R. Norton · J. Szymoniak · J. A. Tunge

Springer

The series *Topics in Organometallic Chemistry* presents critical overviews of research results in organometallic chemistry, where new developments are having a significant influence on such diverse areas as organic synthesis, pharmaceutical research, biology, polymer research and materials science. Thus the scope of coverage includes a broad range of topics of pure and applied organometallic chemistry. Coverage is designed for a broad academic and industrial scientific readership starting at the graduate level, who want to be informed about new developments of progress and trends in this increasingly interdisciplinary field. Where appropriate, theoretical and mechanistic aspects are included in order to help the reader understand the underlying principles involved.

The individual volumes are thematic and the contributions are invited by the volumes editors.

In references Topics in Organometallic Chemistry is abbreviated *Top. Organomet. Chem.* and is cited as a journal

Springer WWW home page: springeronline.com
Visit the TOMC contents at springerlink.com

Library of Congress Control Number: 2004114200

ISSN 1436-6002
ISBN 3-540-22221-9 **Springer Berlin Heidelberg New York**
DOI 10.1007/b80198

Bibliographic information published by Die Deutsche Bibliothek
Die Deutsche Bibliothek lists this publication in the Deutsche Nationalbibliographie; detailed bibliographic data is available in the Internet at <http://dnb.ddb.de>.

This work is subject to copyright. All rights are reserved, whether the whole or part of the material is concerned, specifically the rights of translation, reprinting, reuse of illustrations, recitation, broadcasting, reproduction on microfilm or in any other way, and storage in data banks. Duplication of this publication or parts thereof is permitted only under the provisions of the German Copyright Law of September 9, 1965, in its current version, and permission for use must always be obtained from Springer-Verlag. Violations are liable to prosecution under the German Copyright Law.

Springer is a part of Springer Science+Business Media
springeronline.com
© Springer-Verlag Berlin Heidelberg 2005
Printed in Germany

The use of registered names, trademarks, etc. in this publication does not imply, even in the absence of a specific statement, that such names are exempt from the relevant protective laws and regulations and therefore free for general use.

Typesetting: Fotosatz-Service Köhler GmbH, Würzburg
Production editor: Christiane Messerschmidt, Rheinau
Cover: design & production GmbH, Heidelberg

Printed on acid-free paper 02/3020 me – 5 4 3 2 1 0

Volume Editor

Professor Dr. Ilan Marek
Department of Chemistry
Institute of Catalysis Science and Technology
Technion-Israel Institute of Technology
Haifa 32000
Israel
Chilanm@techunix.technion.ac.il

Editorial Board

Prof. John M. Brown
Dyson Perrins Laboratory
South Parks Road
Oxford OX1 3QY
john.brown@chem.ox.ac.uk

Prof. Pierre H. Dixneuf
Campus de Beaulieu
Université de Rennes 1
Av. du Gl Leclerc
35042 Rennes Cedex, France
Pierre.Dixneuf@univ-rennes1.fr

Prof. Alois Fürstner
Max-Planck-Institut für Kohlenforschung
Kaiser-Wilhelm-Platz 1
45470 Mühlheim an der Ruhr, Germany
fuerstner@mpi-muelheim.mpg.de

Prof. Louis S. Hegedus
Department of Chemistry
Colorado State University
Fort Collins, Colorado 80523-1872, USA
hegedus@lamar.colostate.edu

Prof. Peter Hofmann
Organisch-Chemisches Institut
Universität Heidelberg
Im Neuenheimer Feld 270
69120 Heidelberg, Germany
ph@phindigo.oci.uni-heidelberg.de

Prof. Paul Knochel
Fachbereich Chemie
Ludwig-Maximilians-Universität
Butenandtstr. 5–13
Gebäude F
81377 München, Germany
knoch@cup.uni-muenchen.de

Prof. Gerard van Koten
Department of Metal-Mediated Synthesis
Debye Research Institute
Utrecht University
Padualaan 8
3584 CA Utrecht, The Netherlands
vankoten@xray.chem.ruu.nl

Prof. Shinji Murai
Faculty of Engineering
Department of Applied Chemistry
Osaka University
Yamadaoka 2-1, Suita-shi
Osaka 565, Japan
murai@chem.eng.osaka-u.ac.jp

Prof. Manfred Reetz
Max-Planck-Institut für Kohlenforschung
Kaiser-Wilhelm-Platz 1
45470 Mülheim an der Ruhr, Germany
reetz@mpi.muelheim.mpg.de

**Topics in Organometallic Chemistry
is also Available Electronically**

For all customers who have a standing order to Topics in Organometallic Chemistry, we offer the electronic version via SpringerLink free of charge. Please contact your librarian who can receive a password for free access to the full articles by registration at:

springerlink.com

If you do not have a subscription, you can still view the tables of contents of the volumes and the abstract of each article by going to the SpringerLink Homepage, clicking on "Browse by Online Libraries", then "Chemical Sciences", and finally choose Topics in Organometallic Chemistry.

You will find information about the

– Editorial Board
– Aims and Scope
– Instructions for Authors
– Sample Contribution

at springeronline.com using the search function.

Preface

Over the past few decades, interest in organozirconium chemistry has been rapidly increasing. This special interest arises from the combination of transition metal behavior, such as the coordination of a carbon-carbon multiple bond, oxidative addition, reductive elimination, β-hydride elimination, addition reaction and from the behavior of classical σ-carbanion towards electrophiles. The reactivity of the resulting carbanion can be easily modified by a whole gamut of transmetalation reactions.

In this volume of *Topics in Organometallic Chemistry*, some remarkable recent achievements of organozirconium derivatives are described giving a unique overview of the many possibilities of these organometallic compounds as reagents and catalysts, which is one of the main reasons for their enduring versatility as intermediates over the years.

In this multi-authored monograph, several experts and leaders in the field bring the reader up to date in these various areas of research (synthesis and reactivity of zirconaaziridine derivatives, zirconocene-silene complexes, stereodefined dienyl zirconocenes complexes, octahedral allylic and heteroallylic zirconium complexes as catalysts for the polymerization of olefins and finally the use of zirconocene complexes for the preparation of cyclopropane derivatives). It is their expertise that will familiarize the reader with the essence of the topic.

I wish to express my sincere and deep appreciation to all of them. Reading their contribution was not only a pleasure but will also undoubtedly stimulate future developments in this exciting field of research.

Haifa, September 2004　　　　　　　　　　　　　　　　　　　　　　　　Ilan Marek

Contents

Synthesis and Reactivity of Zirconaaziridines
S. A. Cummings · J. A. Tunge · J. R. Norton 1

Synthesis and Reactivity of Zirconium-Silene Complexes
M. Mori 41

Octahedral Zirconium Complexes as Polymerization Catalysts
A. Lisovskii · M. S. Eisen 63

Zirconocene Complexes as New Reagents for the Synthesis
of Cyclopropanes
J. Szymoniak · P. Bertus 107

Stereoselective Synthesis of Dienyl Zirconocene Complexes
N. Chinkov · I. Marek 133

Author Index Volume 1–10 167

Subject Index 173

Synthesis and Reactivity of Zirconaaziridines

Sarah A. Cummings[1] · Jon A. Tunge[2] · Jack R. Norton[1] (✉)

[1] Department of Chemistry, Columbia University, New York, NY 10027, USA
jrn11@columbia.edu
[2] Department of Chemistry, University of Kansas, Lawrence, KS 66045, USA

1	Introduction	2
2	Synthesis of Zirconaaziridines	2
2.1	From an Amine via C–H Activation by Zr-R	3
2.2	From Coordination of Free Imine	5
2.3	From an Isocyanide via Insertion and Rearrangement	7
2.4	Kinetic vs Thermodynamic Selectivity	8
3	Properties	10
3.1	Structure	10
3.2	Reactivity	14
4	Reactions and Applications	14
4.1	Olefins and Acetylenes	15
4.2	Imines and Aldehydes	21
4.3	Heterocumulenes (CO$_2$, Isocyanates, Carbodiimides)	22
4.3.1	Mechanism of Heterocumulene Insertion; Implication for Other Electrophiles	24
4.4	Carbonates	25
5	Control of Stereochemistry Resulting from Insertion Reactions	26
6	Interconversion of Stereochemistry at Zirconaaziridine Carbons	31
7	Thermal Transformations of Zirconocene-η^3-1-Azaallyl Hydrides	34
8	Conclusions	36
	References	37

Abstract η^2-Imine complexes of zirconium, or zirconaaziridines, have attracted attention as amino carbanion equivalents. They can be prepared from an amine via C–H activation within a zirconium methyl amide complex, from an imine by addition to a zirconocene equivalent, and from an isocyanide by rearrangement of the product of isocyanide insertion into a Zr–H, Zr–C, or Zr–Si bond. Zirconaaziridines contain a chiral center if their ring carbon bears two unequal substituents. Insertion of unsaturated organic compounds into the polar Zr–C bonds of zirconaaziridines leads to amines, allylic amines, heterocycles, diamines, amino alcohols, amino amides, amino amidines, and amino acid esters. These reactions can be carried out asymmetrically if the zirconaaziridines are prepared from enantiomerically pure zirconium complexes, or if racemic zirconaaziridines are treated with enantiomerically pure insertion reagents. Enantioenriched allylic amines and α-amino acid esters have been prepared. The lability of their chiral centers distinguishes zirconaaziridines from other organometallic reagents and makes it possible for them to undergo "dynamic kinetic asymmetric transformations". These chiral centers can invert through isomerization of the zirconaaziridine to (a) (when possible) an azaallyl hydride complex, or (b) (if necessary) an η^1-imine complex. The relative rates of inversion and insertion determine the stereochemical outcome of insertion reactions.

Keywords η^2-Imine complex · Zirconaaziridine · Curtin–Hammett kinetics · Dynamic kinetic asymmetric transformation · Dynamic kinetic resolution

Abbreviations
Cp*   Pentamethylcyclopentadienyl
EBTHI Ethylenebis(tetrahydroindenyl)
L   Ligand

1
Introduction

Metal η^2-imine complexes with various transition metals [1–10] and lanthanides [11, 12] are well known in the literature. Early transition metal η^2-imine complexes have attracted attention as α-amino carbanion equivalents. Zirconium η^2-imine complexes, or zirconaaziridines (the names describe different resonance structures), are readily accessible and have been applied in organic synthesis in view of the umpolung [13] of their carbons; whereas imines readily react with nucleophiles, zirconaaziridines undergo the insertion of electrophilic reagents. Accessible compounds include heterocycles and nitrogen-containing products such as allylic amines, diamines, amino alcohols, amino amides, amino amidines, and amino acid esters. Asymmetric syntheses of allylic amines and α-amino acid esters have even been carried out. The mechanism of such transformations has implications not only for imine complexes, but also for the related aldehyde and ketone complexes [14–16]. The synthesis and properties of zirconaaziridines and their applications toward organic transformations will be discussed in this chapter.

2
Synthesis of Zirconaaziridines

There are three general routes to zirconaaziridines: C–H activation within a zirconium methyl amide complex, imine addition to a zirconocene equivalent, and rearrangement of the product of isocyanide insertion into the Zr–H, Zr–C, or Zr–Si bond. Although a trapping ligand is not required for their generation, zirconaaziridines are often isolated as their 18-electron THF or PMe$_3$ adducts; dissociation of the THF or PMe$_3$ is often required for subsequent reactions. Zirconaaziridines can also be prepared and used in situ in the absence of a coordinating ligand.

2.1
From an Amine via C–H Activation by Zr-R

Zirconocene methyl amide complexes 1 are readily prepared by addition of lithiated secondary or *N*-silyl amines to zirconocene methyl chloride [17] or zirconocene methyl triflate [18] (Eq. 1). Loss of methane from 1 yields zirconaaziridines which, in the presence of THF or PMe$_3$, can be isolated as the adducts 2 in high yield and purity. This synthetic method is ideal when the isolation and characterization of the resulting zirconaaziridine is desired, as the C–H activation and concomitant methane evolution occur with the formation of little side product.

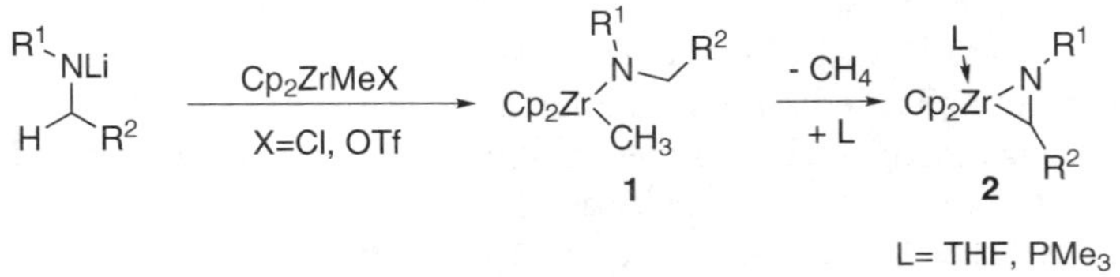

$$\begin{array}{c}R^1\!\!\diagdown\!\!NLi\\H\diagup\!\diagdown\!R^2\end{array} \xrightarrow[X=Cl,\,OTf]{Cp_2ZrMeX} Cp_2Zr\!\diagdown\!\!\begin{array}{c}N(R^1)\!\!-\!\!R^2\\CH_3\end{array} \xrightarrow[+L]{-CH_4} Cp_2Zr\!\diagdown\!\!\begin{array}{c}L\!\!\downarrow\!\!N\!\!-\!\!R^1\\R^2\end{array} \qquad (1)$$

$$\qquad\qquad\qquad\qquad\qquad\qquad\qquad\qquad \mathbf{1}\qquad\qquad\qquad\qquad \mathbf{2}$$

$$L=\text{THF, PMe}_3$$

This method is most practical for complexes where methane elimination is facile and occurs below room temperature. Zirconaaziridines such as those in Fig. 1 (with *N*-aryl and *N*-silyl [19] substituents) have been readily prepared in this manner [20–22]. Buchwald noted that methane elimination from 1 is facile when the "availability of the lone pairs on nitrogen" is reduced [20]; Whitby suggests that such "reduced availability… [is] due to conjugation with the aromatic π system, or overlap with the Si d orbitals (or Si–C σ^* orbitals)" [23]. The rate of zirconaaziridine formation from 1e, for example, is faster by a factor of 1,000 than the rate of the formation from 1f (Fig. 2).

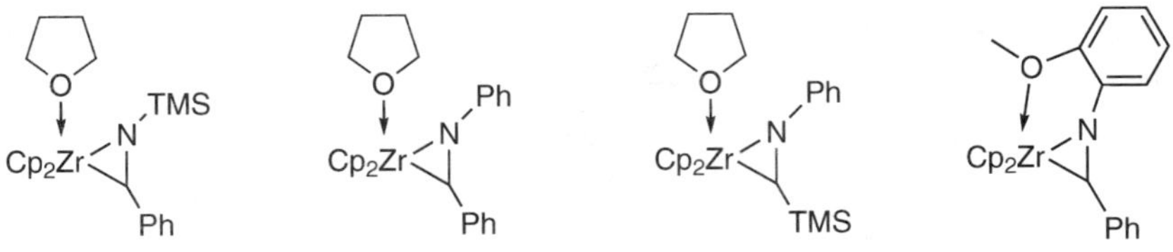

Fig. 1 Examples of zirconaaziridines prepared by C–H activation method

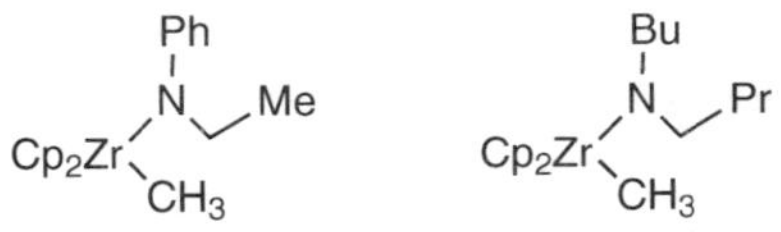

Fig. 2 Rate of methane elimination is 1,000 times faster for 1e than for 1f

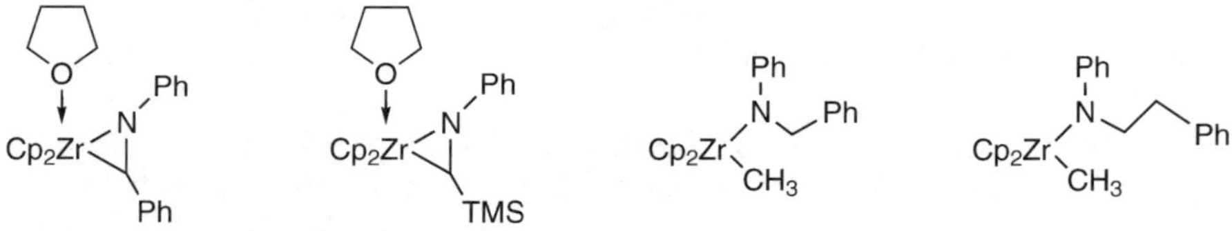

2b **2c** **1b** **1g**

Fig. 3 Carbon substituents that increase the acidity of the methylene proton increase the rate of methane elimination

Substituents in the R^2 position that increase the acidity of the methylene protons on the amide ligand also increase the rate of methane elimination significantly. For example, methane elimination is facile to yield both the phenyl-substituted zirconaaziridine **2b** and the α-silyl-substituted zirconaaziridine **2c** [24], whereas the rate can decrease by a factor of 100 for alkyl-substituted zirconaaziridines [23]. The amide complex **1b** eliminates methane readily below –30 °C, whereas **1g** requires over 1 h at 60 °C (Fig. 3) [24].

These observations, along with kinetic isotope effect studies and Hammett correlation studies, support a concerted elimination by σ-bond metathesis involving a four-membered transition state (Eq. 2) [23]. A large kinetic isotope effect is observed for the loss of methane from methyl amide complexes **1b** and **1h** (Eqs. 3 and 4), comparable to those observed by Buchwald and coworkers for formation of zirconium η^2-thioaldehyde complexes [25] and by Bercaw and coworkers for formation of tantalum η^2-imine complexes [5a] through similar transition states.

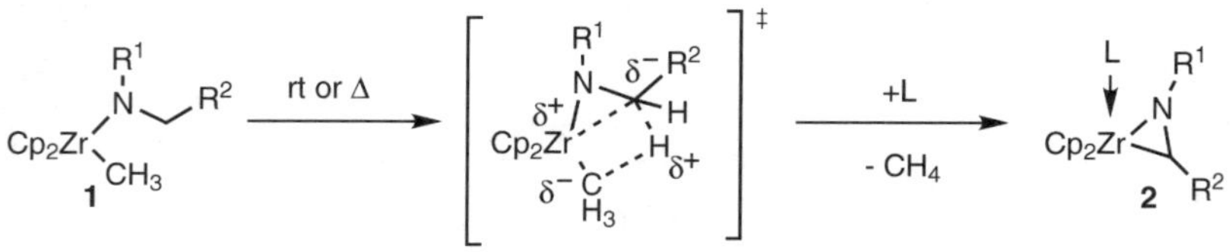

(2)

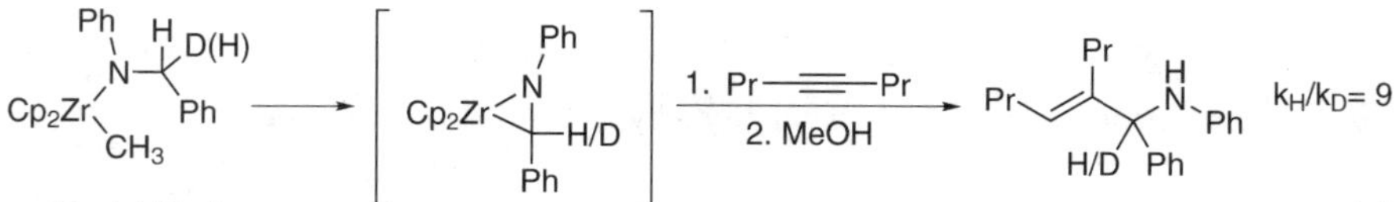

1b-d0/ 1b-d1

(3)

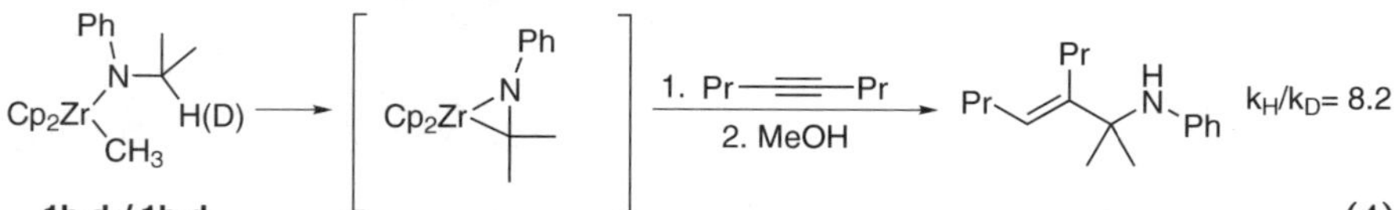

1h-d0/ 1h-d1

(4)

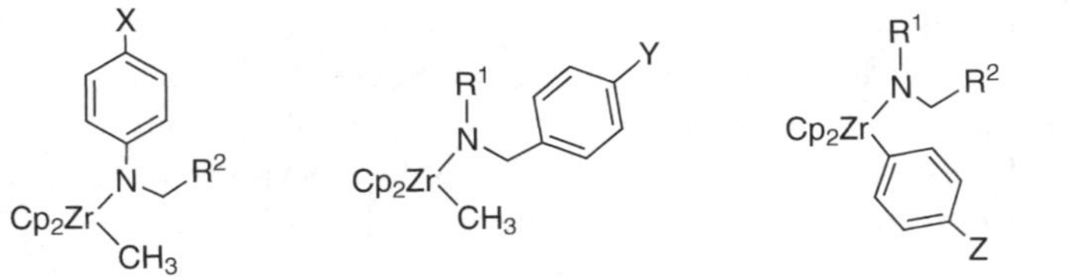

X = CO₂Me, Cl, H, OMe
3

Y = OMe, Cl, H
4

Z = NMe₂, OMe, H, Cl
5

Fig. 4 Compounds used for Hammett plots

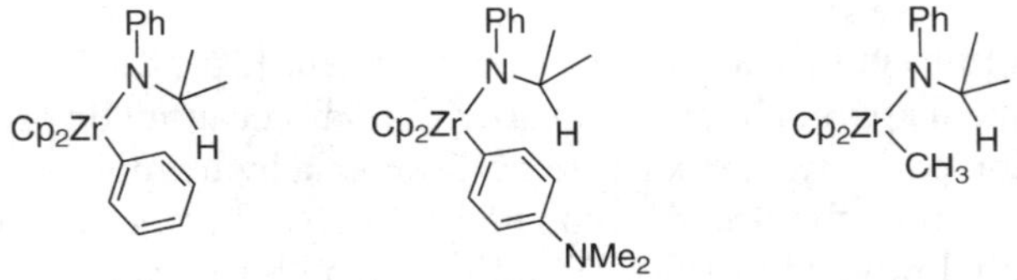

6

7

1h

Fig. 5 Variation of eliminated group

Hammett studies revealed rate acceleration in the presence of electron-withdrawing substituents on **3** (ρ=+3.2). The mechanism requires that the methylene group of the amide ligand and the methyl ligand adopt a geometry in which they can readily interact. Rate acceleration by electron-withdrawing substituents on **4** (ρ=+0.5) and by electron-donating groups on **5** (ρ=–1.56) implies significant negative charge on the methylene carbon and positive charge on zirconium in the transition state. The activation parameter $\Delta S^{\ddagger}$=–4.5 eu agrees with an ordered transition state. All of these results support a polarized four-membered transition state, where the methyl ligand abstracts H⁺ from the methylene group of the amide ligand [23] (Fig. 4).

Zirconaaziridine formation can also occur via benzene elimination [26]. Sequential additions of lithiated amine and phenyllithium to dichlorozirconocene yield a zirconocene phenyl amide complex that undergoes benzene elimination to result in zirconaaziridine. Rates for methane and benzene elimination are similar; for example, the rate of zirconaaziridine formation is not significantly different for complexes **6** and **1h**. However, elimination of a *p*-(dimethylamino)-phenyl group, as in **7**, is a factor of 18 faster than the rate of methane elimination from complex **6**. Thus, altering the group eliminated can facilitate zirconaaziridine synthesis (Fig. 5).

2.2
From Coordination of Free Imine

The synthetic equivalent of zirconocene, "Cp₂Zr" [27], can react with an imine to yield a zirconaaziridine (Eq. 5) [28] – a reaction that reminds us that zirconaaziridines can be seen as η^2-imine complexes of zirconium(II) (see Sect. 3.1). "Cp$_2$Zr", the Negishi reagent, is more complicated than its formula suggests. It is generated by the addition of two equivalents of n-butyllithium to zirconocene dichloride, but the nature of the active species is still not clear; it has attracted attention and debate in the literature [27, 29, 30].

$$\text{Cp}_2\text{ZrCl}_2 \xrightarrow[-78^\circ\text{C}]{2\,\text{BuLi}} \quad \underset{\text{rt}}{\xrightarrow{\text{H}\overset{\text{N-R}^1}{\underset{}{\|}}\text{R}^2 \,,\, \text{L}}} \quad \underset{\mathbf{2}}{\text{Cp}_2\text{Zr}\!\!\begin{array}{c}\text{L}\\|\\\text{N-R}^1\\|\\\text{R}^2\end{array}} \quad (5)$$

Zirconaaziridine synthesis via the Negishi reagent [22, 24, 28, 31, 32] is a particularly efficient approach for zirconaaziridines that cannot be prepared by the C–H activation pathway. For example, addition of a hydrazone to "zirconocene" yields an N-amino-substituted zirconaaziridine (see Sect. 4.1). The ligand-free **2i**, produced in low yield by the C–H activation pathway, can be prepared from "zirconocene" and an imine/enamine mix as shown in Eq. 6 [24].

$$\text{Cp}_2\text{ZrCl}_2 \xrightarrow[-78^\circ\text{C}]{2\,\text{nBuLi}} \xrightarrow[\text{rt}]{\underset{\text{Me}}{\overset{\text{N-TMS}}{\|}}\text{Ph} \,+\, \underset{\text{H}_2\text{C}}{\overset{\text{HN-TMS}}{\|}}\text{Ph}} \underset{\mathbf{2i}}{\text{Cp}_2\text{Zr}\!\!\begin{array}{c}\text{N-TMS}\\|\\\text{Ph}\\|\\\text{Me}\end{array}} \quad (6)$$

With its tetrasubstituted carbon, the zirconaaziridine **2i** remains ligand-free in THF, but in the presence of PMe$_3$ it forms a PMe$_3$ adduct. Complex **2j** [33], prepared in a similar manner with the Rosenthal reagent **8** [34], is the only zirconaaziridine with a tetrasubstituted carbon that has had its structure determined by X-ray studies. The stability of its pyridine adduct reflects the fact that pyridine is a stronger donor ligand than THF.

$$\text{Cp}_2\text{ZrCl}_2 \xrightarrow[\substack{\text{Mg}\\\text{THF}}]{\text{TMS}\!\equiv\!\text{TMS}} \underset{\mathbf{8}}{\text{Cp}_2\text{Zr}\!\!\begin{array}{c}\text{THF}\\|\\\text{TMS}\\\\\text{TMS}\end{array}} \xrightarrow{\underset{\text{Me}}{\overset{\text{N-Ph}}{\|}}\text{Ph},\ \text{pyridine}} \underset{\mathbf{2j}}{\text{Cp}_2\text{Zr}\!\!\begin{array}{c}\text{pyridine}\\|\\\text{N-Ph}\\|\\\text{Ph}\\|\\\text{Me}\end{array}} \quad (7)$$

Zirconaaziridines prepared through the addition of imine to "zirconocene" are generally not of high purity and cannot be isolated in high yield; reductive coupling of imines often competes. However, this method is preferred for preparing zirconaaziridines in situ for reaction with electrophiles that can be used in excess, limiting the amount of imine self-coupling. The fact that the Negishi reagent can be prepared in a single step from "bench-stable" materials is a great advantage.

2.3
From an Isocyanide via Insertion and Rearrangement

A zirconaaziridine can also result from the thermal rearrangement of an iminoacyl hydride complex, formed by isocyanide insertion into one Zr–H bond of a dihydride. Bercaw and coworkers observed (Eq. 8) that ArNC insertion into a Zr–H bond of $Cp^*_2ZrH_2$, followed by hydride transfer to the iminoacyl ligand of **9**, formed the zirconaaziridine **10**, although **10** rearranged upon further thermolysis [35].

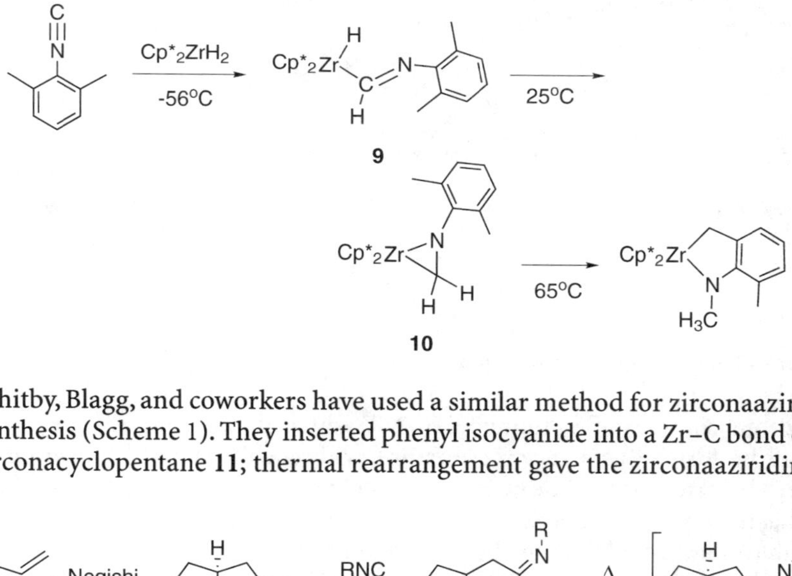

(8)

Whitby, Blagg, and coworkers have used a similar method for zirconaaziridine synthesis (Scheme 1). They inserted phenyl isocyanide into a Zr–C bond of the zirconacyclopentane **11**; thermal rearrangement gave the zirconaaziridine **12**.

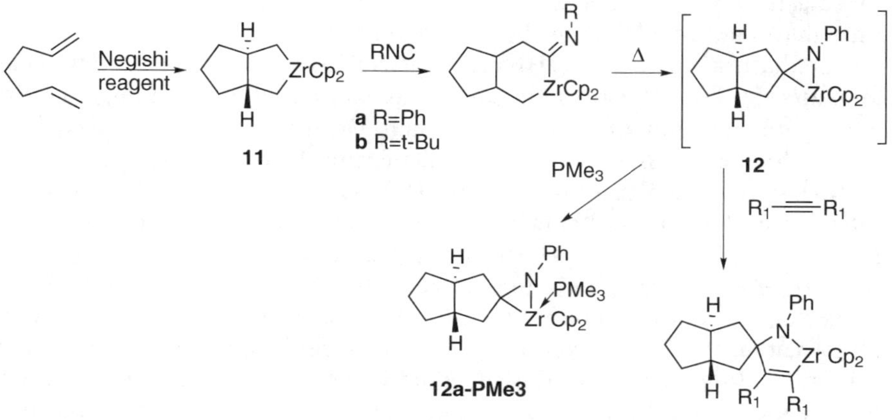

Scheme 1 Formation and trapping of zirconaaziridine after isocyanide insertion into Zr-R

Though not general (rearrangement is not observed for R=*t*-Bu, perhaps because its steric bulk prohibits coordination of the nitrogen), this method allows the synthesis of the spirocyclic zirconaaziridine **12** [36, 37].

In a related method, reduction of an iminosilaacyl ligand on zirconium also yields a zirconaaziridine. The Zr–Si bond in complex **13** is known to insert carbon monoxide or isocyanides to yield a silaacyl or iminosilaacyl complex such as **14** [38]. The air- and moisture-stable **14** has been studied by X-ray crystallography [39]. Mori and coworkers have reported that treating complexes **14** with LiEt$_3$BH results in replacement of the chloride by hydride; rearrangement yields the silyl-substituted zirconaaziridine **2k**. Trapping **2k** with 4-octyne followed by aqueous workup yields the α-silyl allylic amine **15** [39–41].

$$\text{(9)}$$

2.4
Kinetic vs Thermodynamic Selectivity

The formal complex of an imine RCH=NR' with a zirconocene of C_{2v} symmetry will be formed as a racemic mixture, with equal amounts of two enantiomers. If a zirconaaziridine possesses another stereogenic element, it will have two diastereomers of unequal energy, and they may be generated in a kinetic ratio that is not equal to their thermodynamic ratio.

The diastereomers of EBTHI zirconaaziridines are formed in comparable amounts via C–H activation, but equilibrate quickly (within an hour) to yield a thermodynamic mixture of diastereomers. Grossman observed little difference in the loss of MeH vs MeD from deuterium-labeled (EBTHI)zirconium methyl amide complexes **16l** (Scheme 2). Loss of MeH and of MeD lead to different diastereomers, although their relative rate will also reflect the primary kinetic isotope effect for C–H activation. Neither k_1/k_2 nor k_3/k_4 is large, and both are largely the result of isotope effects rather than diastereoselectivity [42].

Norton and coworkers noticed significant differences in diastereoselectivity for formation of EBTHI zirconaaziridines **17m** under kinetic vs thermodynamic conditions. In general, **17m** was prepared by heating the (EBTHI)zirconium methyl amide complex at 70 °C (Scheme 3). Use of a 20-fold excess of carbonate, by accelerating insertion relative to diastereomer equilibration, permitted

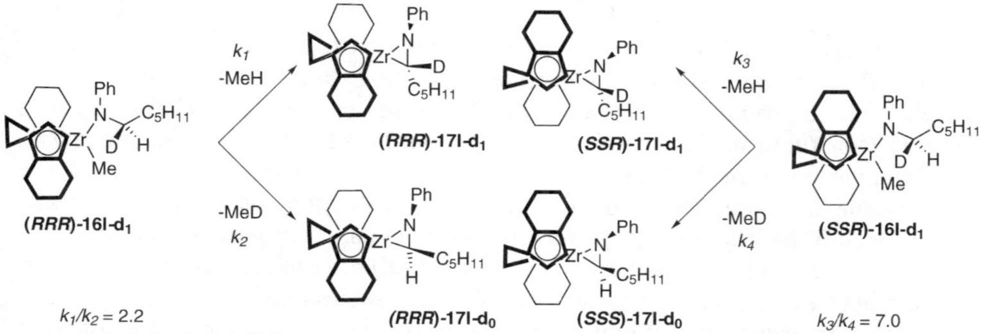

Scheme 2 Deuterium labeling experiment to determine diastereoselectivity in zirconaaziridine formation

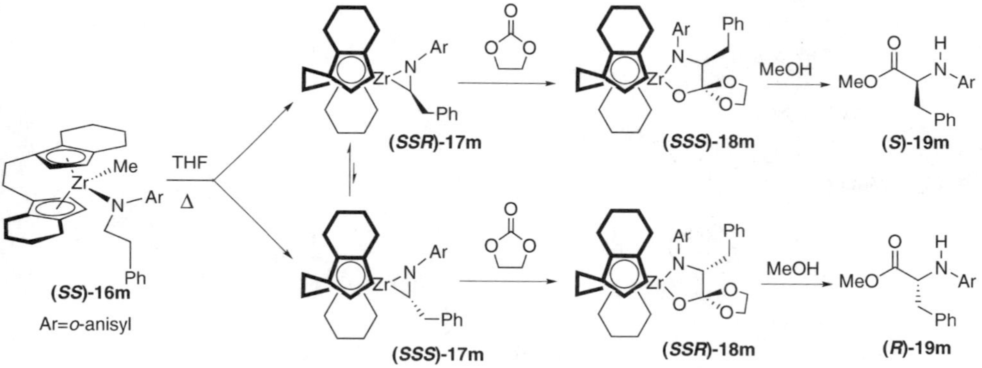

Scheme 3 Origin of diastereoselectivity under thermodynamic and kinetic conditions

the kinetic ratio to be trapped. Under these conditions, 54% ee of the (*R*)-amino acid ester **19m** was observed. Insertion appears to occur with retention of configuration, so the 54% ee of the (*R*)-**19m** reflects the de of the insertion products (*SSR*)-**18m** and (*SSS*)-**17m** present under kinetic conditions. (The CIP stereochemical descriptor changes upon insertion.) Allowing the zirconaaziridine to equilibrate at room temperature before addition of ethylene carbonate (EC) gave a thermodynamic distribution of diastereomers. The 85% ee of the (*S*)-amino acid ester **19m** eventually obtained reflects the de of the (*SSR*)-**17m** present under thermodynamic conditions. The thermodynamically favored zirconaaziridine is that with its carbon substituent oriented away from the six-membered ring of the EBTHI ligand, (*SSR*)-**17m**. Thus, different diastereomer ratios (*SSR*)-**17m**/(*SSS*)-**17m** are observed under thermodynamic and kinetic conditions [43]. Details regarding the reaction of **17m** and ethylene carbonate at various concentrations and the mechanistic implications of the results will be forthcoming in Sect. 5.

Similarly, a difference between the kinetic and thermodynamic selectivities was observed by Taguchi and coworkers while generating zirconaaziridines by addition of the optically active imine to "zirconocene" [31] (Scheme 4). The Negishi reagent was treated with the imine at 0, 23, and 67 °C. The zirconaaziridine diastereomer ratio was determined from the *RRR:SRR* ratio of the amino alcohols resulting from benzaldehyde insertion, *on the assumption that the results of the insertion reaction reflect the ratio of the zirconaaziridine diastereomers*. For **2n** the diastereomer ratio was high (96:4) at 0 °C, but in the opposite direction (5:95) at 67 °C. The diastereomer ratio observed at 0 °C was apparently kinetic, resulting from the relative ease of approach to "zirconocene" of the inequivalent faces of the aldimine. Equilibration of the diastereomers at higher temperature gave the thermodynamic ratio. Taguchi and coworkers proposed that diastereomer interconversion at high temperatures occurred by dissociation and recomplexation of the imine to "zirconocene". The factors that control product stereochemistry in this system will be discussed in Sect. 5.

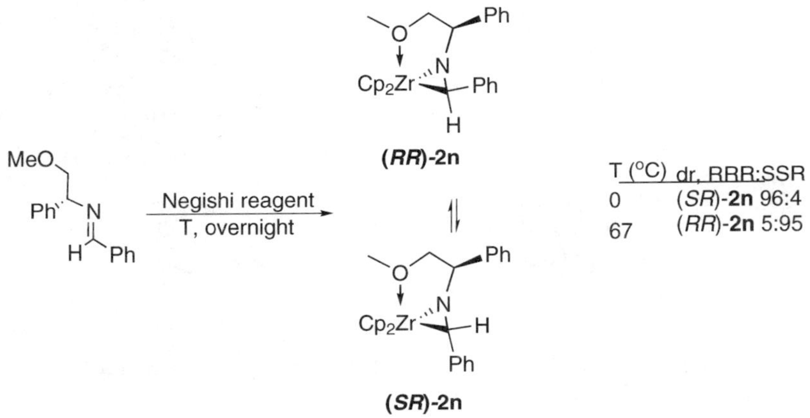

Scheme 4 Diastereomer ratio of zirconaaziridine and amino alcohol product of aldehyde insertion under thermodynamic and kinetic conditions for zirconaaziridine formation from the Negishi reagent

3
Properties

3.1
Structure

The two extremes of the Dewar–Chatt–Duncanson model for olefin coordination can also be applied to describe aldehyde, ketone, and imine complexes. Resonance structure **A** is an η^2 complex of Zr(II), while its resonance structure **B** is a zirconaaziridine containing Zr–C and Zr–N bonds (Fig. 6). X-ray structural studies of zirconaaziridines and their observed reactivity suggest that resonance structure **B** is more important.

Synthesis and Reactivity of Zirconaaziridines

$$\text{Zr} \leftarrow \overset{N}{\underset{\|}{\|}} \qquad \text{Zr} \overset{N}{\underset{}{<}}|$$

  A   B

Fig. 6 Resonance structures for η^2-imine complexes of zirconium

X-ray structural studies have been conducted on the zirconaaziridines **2a** [20], **2b** [21], **2d** [44], and **2j** [33] (Fig. 7). Relevant data are listed in Table 1. In each case, the stabilizing donor ligand such as THF, *o*-anisyl, or pyridine (referred to as L in the table) prefers to coordinate on the nitrogen side of the zirconaaziridine (Fig. 8), with implications for reactivity that will be discussed in Sect. 4.3.1. Evidence that such a preference is thermodynamic can be seen in the constant 2.3:1 "inside":"outside" [45] ratio for zirconaaziridine **2o** regardless of the method used for its generation (Fig. 8) [32]. This inside coordination is also known to be thermodynamically favored over outside coordination for most acyl [14] and iminoacyl [32] complexes. The *o*-anisyl oxygen is clearly seen to coordinate on the inside in the structure of **2d** (Fig. 9).

The zirconaaziridines **2a**, **2b**, **2d**, and **2j** show significant elongation of the C–N bond length (Table 1, entry 1) compared to that of a free imine (1.279 Å) [46], adopting a bond length much closer to that of a free *amine* (1.468 Å) [46] and confirming that resonance structure **B** (Fig. 6) is important. The Zr–C bond

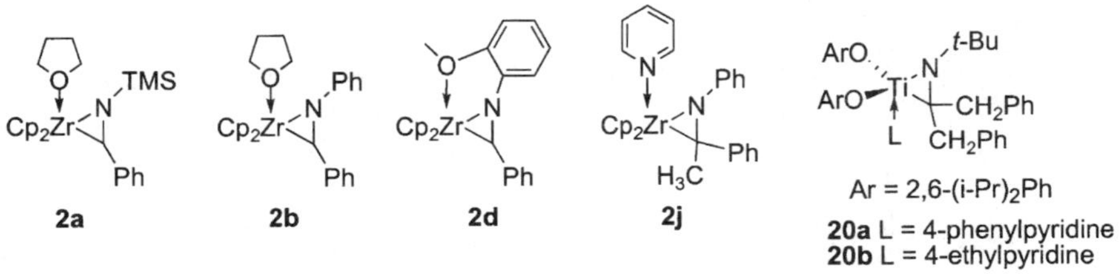

Fig. 7 Zircona- and titanaaziridines studied by X-ray crystallography

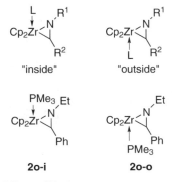

Fig. 8 Ligands prefer "inside" coordination

Table 1 Representative X-ray data on zircona- and titanaaziridines

Entry	Compound	2a	2b	2d	2j	20a	20b
1	C–N (Å)	1.41(1)	1.431(7)	1.402(5)	1.436(5)	1.421(7)	1.410(3)
2	Zr–C (M–C)	2.26(1)	2.299(5)	2.374(3)	2.364(4)	2.158(5)	2.150(2)
3	Zr–N (M–N)	2.11(1)	2.113(4)	2.042(3)	2.103(3)	1.846(4)	1.855(2)
4	Zr–L (M–L)	2.376(9)	2.340(4)	2.436(3)	2.415(4)	2.216(5)	2.224(2)
5	C–C(Ph)	1.48(2)	1.465(8)	1.473(5)	1.483(6)		
6	N–Ar/Si	1.69(1)	1.375(7)	1.356(4)	1.388(5)		
7	N–Zr–C (°)	37.5(4)	37.6(2)	36.0(1)	36.92(13)	40.6(2)	40.41(8)
8	N–Zr–L	80.7(3)	80.7(2)	65.9(1)	81.42(12)	130.5(2)	129.76(7)
	N–M–L						
9	N–C–Zr	65.5(6)	64.1(3)	59.0(2)	61.6(2)		
10	Zr–C–C(Ph)	128.6(8)	126.1(4)	128.0(2)	119.9(3)		
11	C–N–Ar/Si	125.3(8)	121.1(4)	136.2(3)	122.7(3)		
12	C–N–Zr	77.0(7)	78.3(3)	85.0(2)	81.5(2)		
13	Zr–N–Ar/Si	148.1(6)	142.2(4)	130.8(2)	150.3(3)		

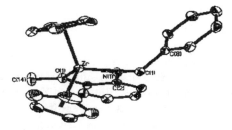

Fig. 9 X-ray crystal structure of **2d**, showing "inside" coordination of methoxy group

lengths (entry 2) for each zirconaaziridine can be compared to the Zr(IV)-alkyl bond lengths (average 2.277 Å) observed for Cp$_2$ZrMe$_2$ [47]. The longer Zr-C bond length observed in **2d** can be attributed to the effect of the chelate ring, whereas the longer Zr-C bond in **2j** reflects the increased congestion at the zirconaaziridine carbon.

While there have been no X-ray structure studies on η^2-imine complexes of hafnium, there are several on η^2-imine complexes of titanium [1c,d]; these allow comparison between Group 4 metallaaziridines. As expected, the Ti-C and Ti-N bonds are shorter than the analogous Zr-C and Zr-N bonds. The titanaaziridines **20a** and **20b** also appear to have a C-N bond order of one (entry 2).

Imine, aldehyde, and ketone complexes, unlike olefin complexes, can also adopt an η^1 configuration (Fig. 10). To our knowledge, η^1 coordination of an imine has never been confirmed by X-ray crystallography, but there are persuasive arguments for such a structure in the ^{13}C NMR spectra of two Zr imine complexes. The ^{13}C NMR resonances for the imine carbons in **21** [43] and **22** [48] occur far downfield from those observed for zirconaaziridines [20] (Table 2).

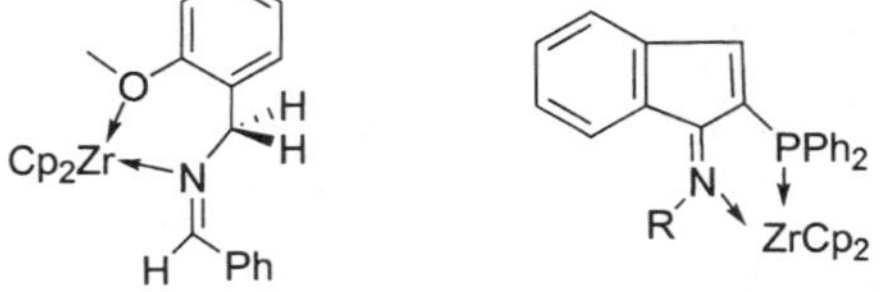

Fig. 10 Zr η^1-imine complexes

Table 2 ^{13}C chemical shifts of the carbons in zirconaaziridines and η^1-imine complexes

Complex	2a	2a-PMe$_3$	2b-PMe$_3$	21	22a	22b
δ^{13}C	68.26	56.1	46.81	176	118.9	136.2

3.2
Reactivity

The structural differences between η^1- and η^2-imine complexes result in differences in their reactivities. While η^1-imine complexes do not undergo coupling reactions with electrophiles, titana- and zirconaaziridines react like d[0] Ti/Zr(IV) organometallic complexes; they readily insert multiple bonds and undergo coupling reactions. Although extremely air- and moisture-sensitive, zirconaaziridines are stable indefinitely under inert atmosphere. Zirconaaziridines can also be generated in situ for use in organic synthesis.

4
Reactions and Applications

Zirconaaziridines react with unsaturated C–C bonds such as (1) olefins and acetylenes [20], and with unsaturated C–X bonds such as (2) aldehydes and imines [20], (3) heterocumulenes [21, 43, 49], and (4) carbonates [21, 22, 43, 50] (Scheme 5). The products generated upon workup are α-functionalized amines. Asymmetric transformations can be carried out when a chiral zirconaaziridine or inserting reagent is used; optically active allylic amines and amino acid esters have been prepared, and the details of these transformations will be discussed.

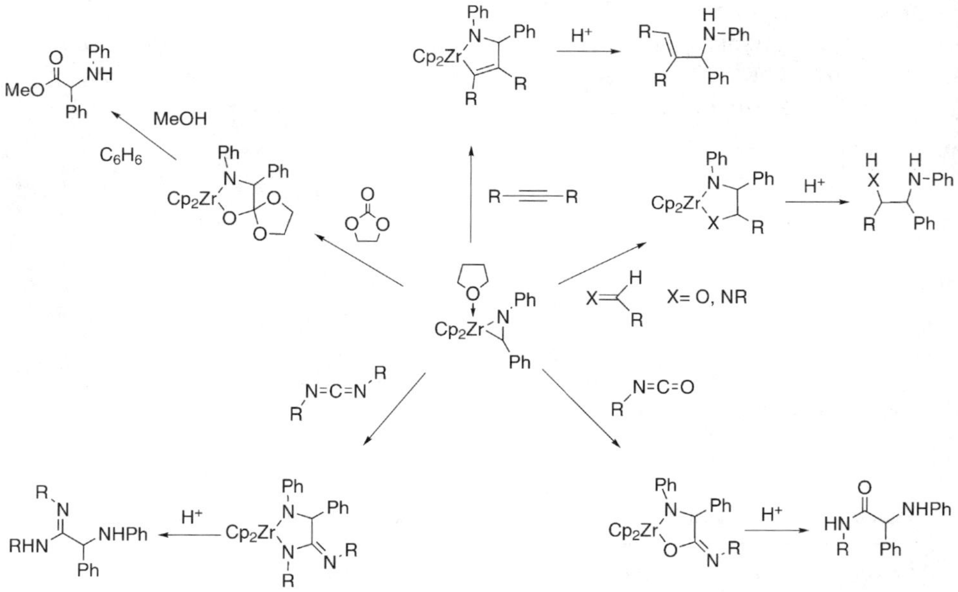

Scheme 5 Zirconaaziridines undergo insertion of various unsaturated compounds into their Zr–C bonds

4.1
Olefins and Acetylenes

Intermolecular olefin and acetylene insertion into zirconaaziridines has been studied by Buchwald and coworkers. High regio- and diastereoselectivity is observed for the formation of the 3,4-disubstituted product 23 in Eq. 10 [20]. Only a single diastereomer of the resulting chiral amine is isolated upon acidic workup.

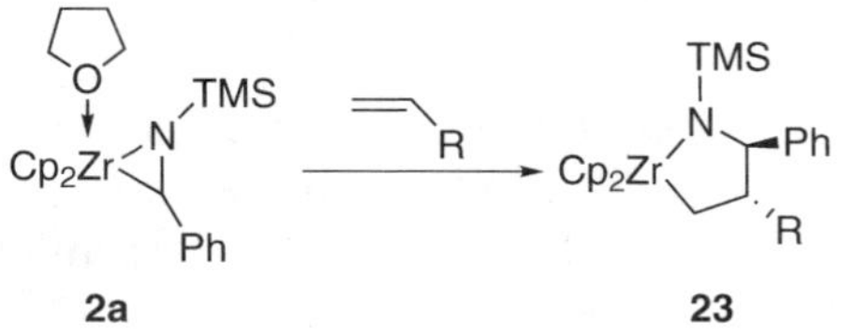

R	%	de
H	62	n/a
Me	88	100
nBu	96	100

(10)

Significant regioselectivity is also observed for the insertion of unactivated alkynes into the Zr–C bonds of zirconaaziridines. With terminal alkynes, the regioselectivity is determined by the substituents on the zirconaaziridine and the alkyne. If R is aryl, the only product is 24. Alkyl substituents, as shown in Scheme 6, decrease the selectivity for 24 over 25 [20].

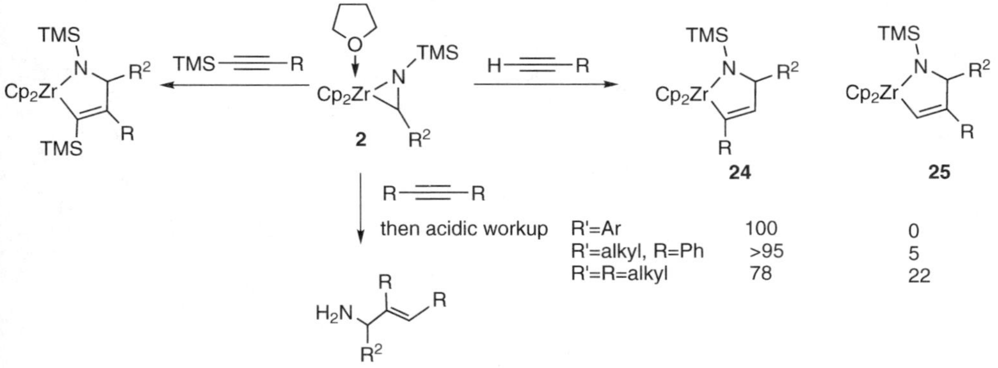

	24	25
R'=Ar	100	0
R'=alkyl, R=Ph	>95	5
R'=R=alkyl	78	22

Scheme 6 Alkyne insertion into zirconaaziridines

The eventual product of such reactions (after cleavage of the Zr–C bond during workup [51]) is an amine with a vinyl or alkyl substituent (arising from the alkyne or alkene) at the α position. Whitby thus derivatized tetrahydroquinoline, following in situ generation of the zirconaaziridine 26, by alkyne, alkene, or allene insertion (Eq. 11) [52, 53].

N-heterocycles can be prepared by inserting olefins or acetylenes with pendant electrophiles. Typical electrophiles used are alkyl halides and epoxides. The latter do not react with the zirconaaziridine, but with the amine generated during workup, resulting in cyclization. The diastereopure amines below cyclize to yield pyrrolidines ($n=1$), piperidines ($n=2$), and perhydroazepines ($n=3$) (Eqs. 12 and 13). Whitby applied this method to the convergent total synthesis of the tetrahydroisoquinoline **27** (Scheme 7) [54].

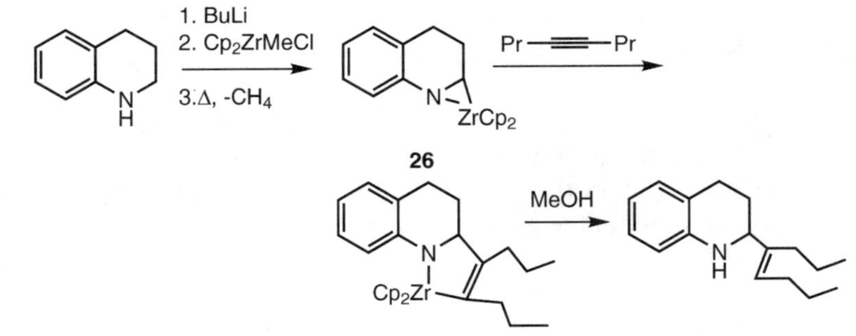

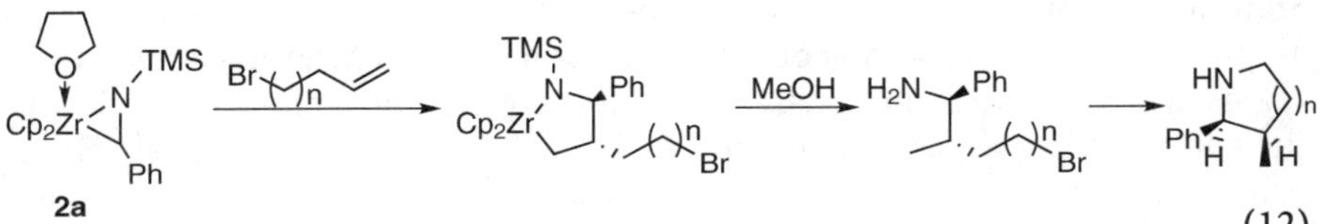

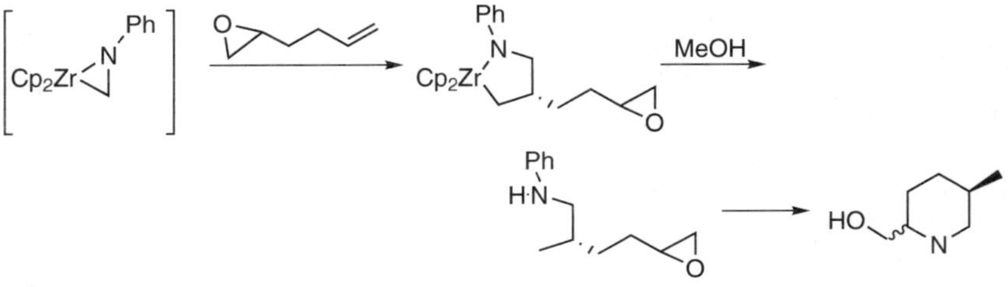

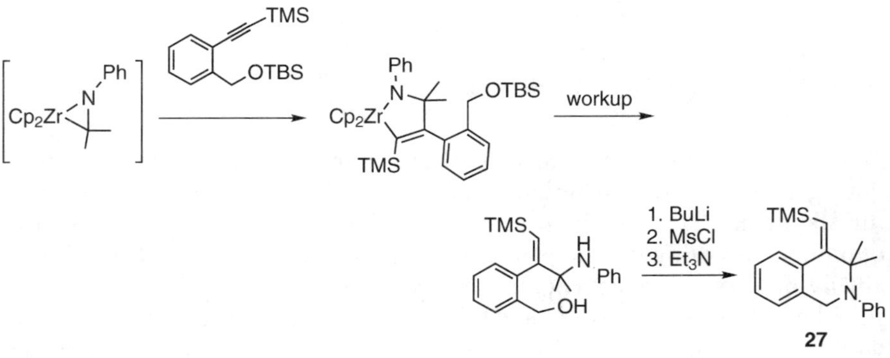

Scheme 7 Application of zirconaaziridines to the synthesis of a tetrahydroisoquinoline

Synthesis and Reactivity of Zirconaaziridines

Similarly, the regio- and diastereoselective formation of azetidines can be accomplished through olefin insertion into the Zr–C bond of zirconaaziridines (Eq. 14). Cleavage of the Zr–C bond with I_2 introduces an alkyl iodide functionality, which alkylates at nitrogen to result in intramolecular cyclization and the diastereoselective formation of azetidines, the products of the formal zirconium-mediated [2+2] reaction of an imine and an olefin. Hindered cyclic alkenes can also insert into the Zr–C bonds of zirconaaziridines to yield bicyclic products [55], albeit with low diastereoselectivity (2:1 for norbornene).

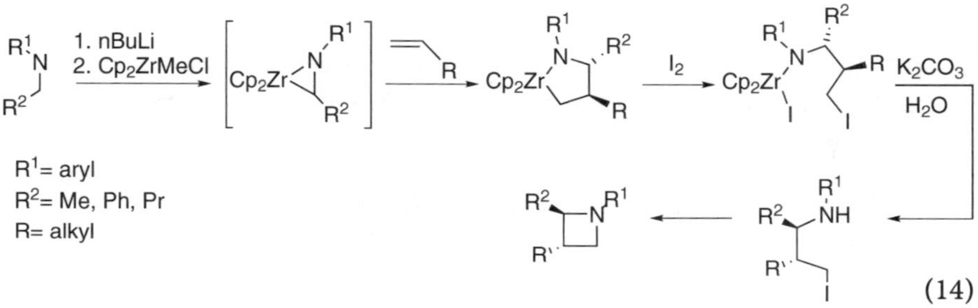

R^1 = aryl
R^2 = Me, Ph, Pr
R = alkyl

(14)

Asymmetric versions of these transformations can be carried out when resolved EBTHI zirconaaziridines are used (Eq. 15). The allylic amine products (Table 3) resulting from alkyne insertion followed by workup are not only geometrically pure, but also enantiomerically enriched; the ee is high for a variety of substrates [18]. A phenyl substituent (R^1) on the zirconaaziridine

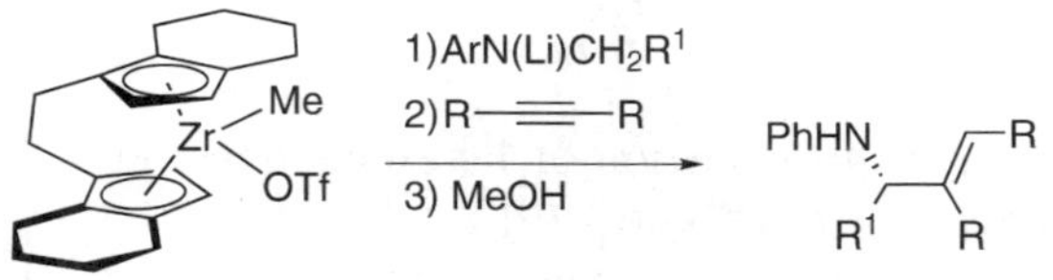

(15)

Table 3 Products, yield, and selectivity for alkyne/alkene insertion into EBTHI zirconaaziridines

Entry	Product	% yield	% ee	de or regioselectivity
1	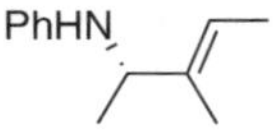 PhHN	72	>95	
2	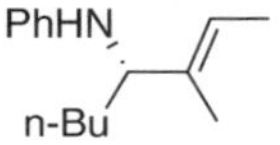 PhHN, n-Bu	72	>95	
3	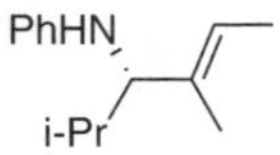 PhHN, i-Pr	60	>95	

Table 3 (continued)

Entry	Product	% yield	% ee	de or regioselectivity
4	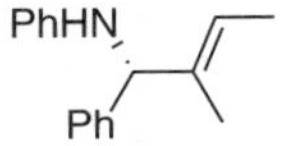 PhHN–/Ph (vinyl)	38	18	
5	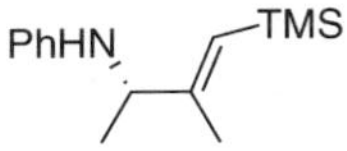 PhHN–/TMS	68	>90	24:1
6	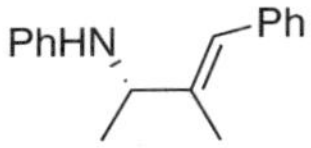 PhHN–/Ph	50	99	100:0
7	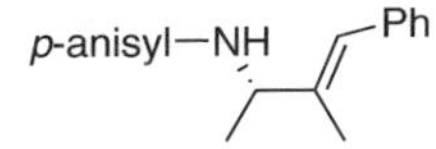 p-anisyl–NH–/Ph	64	>95	
8	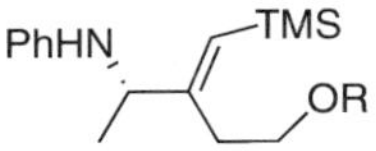 PhHN–/TMS, OR	59	>95	17:1
9	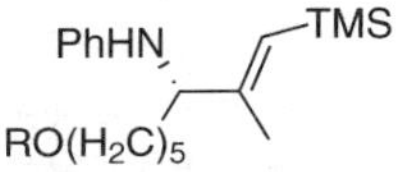 PhHN–/TMS, RO(H₂C)₅	53	>95	22:1
10	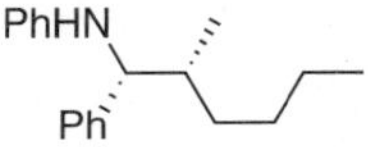 PhHN–/Ph	54	High (single zirconacycle)	8:1

carbon resulted in a low ee (entry 4); insertion of 1-hexene, however, occurs selectively into the same zirconaaziridine (entry 10).

Intramolecular insertion of olefins and acetylenes into the Zr–C bonds of zirconaaziridines can lead to annulation reactions. Livinghouse showed that alkyl-, silyl-, and aryl-substituted C–C multiple bonds readily insert into the Zr–C bonds of zirconaaziridines derived from the hydrazones **28** and **29**, yielding cyclic products [28].

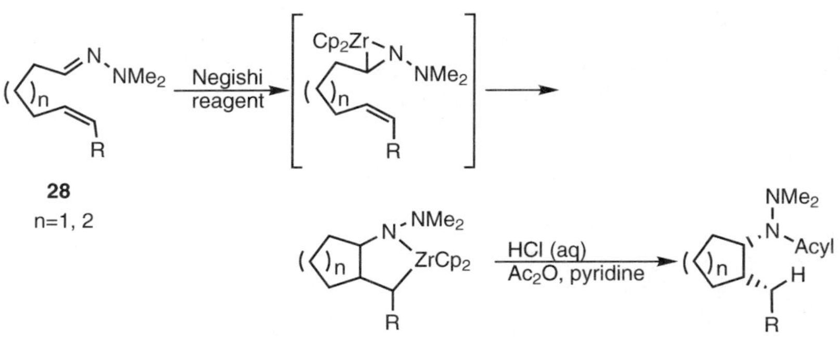

$$\text{(16)}$$

28
n=1, 2

Synthesis and Reactivity of Zirconaaziridines

$$\text{29} \xrightarrow{\text{Negishi reagent}} \left[\text{Cp}_2\text{Zr intermediate} \right] \longrightarrow$$

(29): n=1, 2; R=H, Bu, TMS, Ph

$$\xrightarrow{\text{HCl (aq)}}_{\text{Ac}_2\text{O, pyridine}} \quad (17)$$

Further transformations of such zirconacycles are possible before the organic product is removed from the metal center. Buchwald followed the insertion of an alkyne into a zirconaaziridine with the insertion of CO to obtain substituted pyrroles (in a single pot at ambient temperature) in moderate to good yields. A diverse array of substituents can be introduced by varying the amine and alkyne; examples are shown in Table 4. The reaction is tolerant not only of various alkyls and aryls, but also of thiophenyl, furyl, and pyrrole substituents, and can create 2,2′-bipyrroles (see Table 4). It does, however, require high CO pressure (as much as 1,500 psi can be necessary) [56].

$$R^1 \equiv\!\!\equiv R^2 + CO + \text{TMS-N(R)-CH}_2 \xrightarrow{[\text{Zr}]} \text{pyrrole} \quad (18)$$

Table 4 Substrates and yields for Zr-mediated pyrrole synthesis

R	R^1	R^2	Yield (%)
Ph	H	Ph	69
Ph	H	H	41
Ph	Ph	Ph	49
Ph	Me	Me	44
Ph	H	CH_2OTBS	31
Ph	H	$(CH_2)_3CN$	48
Ph	H	Pr	56
H	n-C_8H_{17}	H	40
n-C_5H_{11}	H	Ph	54
2-thiophenyl	H	Pr	41
2-furyl	H	Ph	49
2-(4-methyl)pyrrole	H	Pr	68

A proposed mechanism is shown in Scheme 8. After alkyne insertion generates an azazirconacyclopentene, CO insertion in the Zr–C(sp^2) bond gives a six-membered ring. Rearrangement of the insertion product forms a zircona-oxirane [14, 15]. Cleavage of the C–O bond, protonolysis of the Zr–C bond, and aromatization of the pyrrole ring then yields the functionalized pyrrole [56]. Classical pyrrole syntheses, such as the Paal–Knorr reaction [57], require the condensation of carbonyl compounds with amines, often via several steps at high temperatures. The fact that Zr is inexpensive makes it possible to consider stoichiometric approaches like Eq. 18, although Arndtsen [58] has just published a Pd-catalyzed synthesis of pyrroles from imines, alkynes, and acid chlorides.

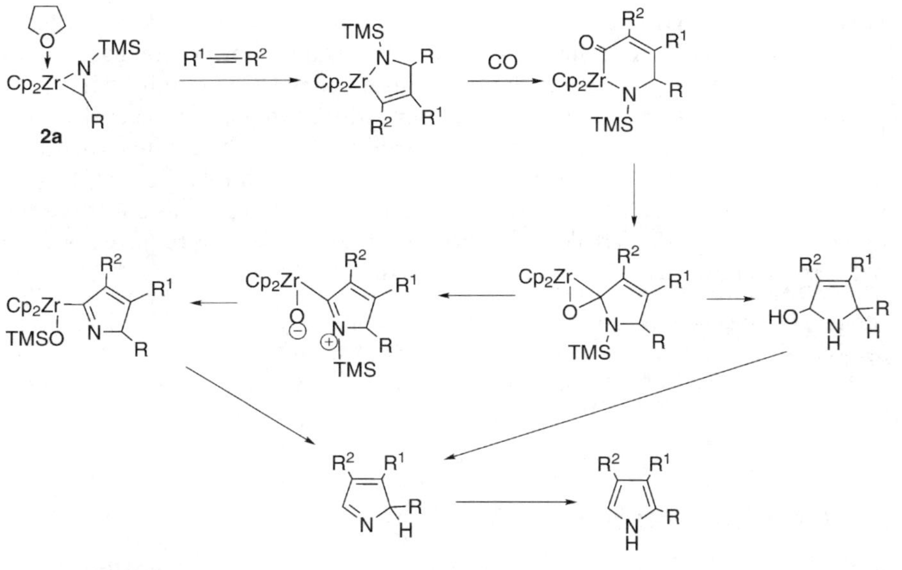

Scheme 8 Proposed mechanism for pyrrole synthesis from an amine, alkyne, and CO via zirconaaziridine

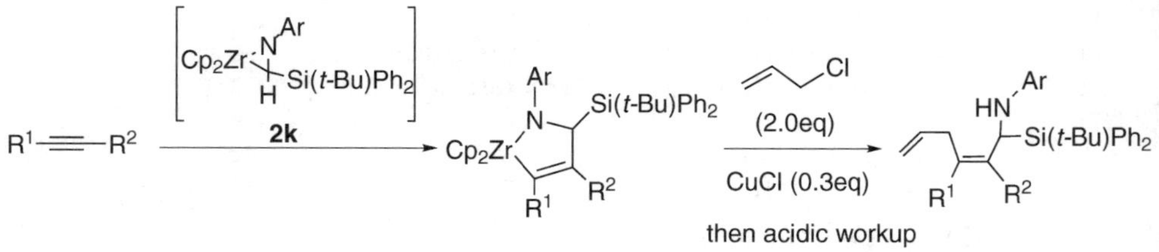

R^1= Ph	R^2= nC_5H_{11}	87%
TMS	Me	87
BR_2	nC_5H_{11}	85
$(CH_2)_3OBn$	$(CH_2)_3OBn$	53

(19)

Mori has employed transmetallation, well documented in other organozirconium reactions [59], to form γ-allylated α-silyl allyl amines (Eq. 19) [40, 41] and substituted pyrroles (Eq. 20) [60] after the insertion of alkynes into the Zr–C bond of zirconaaziridines. The reaction sequence in Eq. 19 uses stoichiometric quantities of Zr, but catalytic amounts of Cu, and both the insertion and allylation steps proceed with high regioselectivity. Using an acyl halide instead of allyl chloride gives tetra- and pentasubstituted pyrroles, in a one-pot reaction (Eq. 20).

$$(20)$$

4.2
Imines and Aldehydes

An imine can insert into the Zr–C bond of a zirconaaziridine resulting in the reductive coupling of the two imines. Reaction of **2a** with TMS-benzaldimine yields principally the *anti* product in high yield (Eq. 21). Aldehydes also insert regioselectively into the Zr–C bond of zirconaaziridines, resulting after workup in amino alcohols. The reaction of the zirconaaziridine **2a** with benzaldehyde or 1,1,1-trimethylacetaldehyde generated the products of reductive coupling in 6.4:1 and >99:1 ratios of *trans:cis*, respectively [20]. Treating the imine and alde-

$$(21)$$

hyde insertion products with ethyl chloroformate resulted in imidazolidinone **30** and oxazolidinone **31**, respectively (Eqs. 21 and 22) [42].

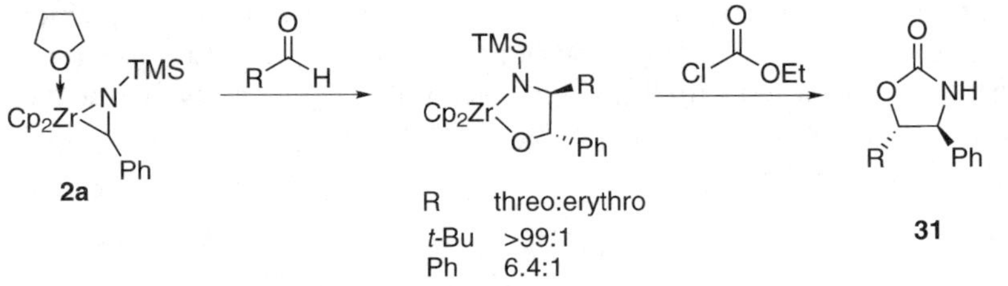

	threo:erythro
t-Bu	>99:1
Ph	6.4:1

(22)

4.3
Heterocumulenes (CO$_2$, Isocyanates, Carbodiimides)

Carboxylation of an amine should be possible by insertion of CO$_2$ into a Zr-C bond of a zirconaaziridine. However, treatment of **2a** and **2b** with CO$_2$ (Eq. 23) gave intractable white precipitates [21, 61]. Treatment of **2d** with CO$_2$ gave a stable insertion product which, however, proved difficult to cleave from zirconium (Eq. 24).

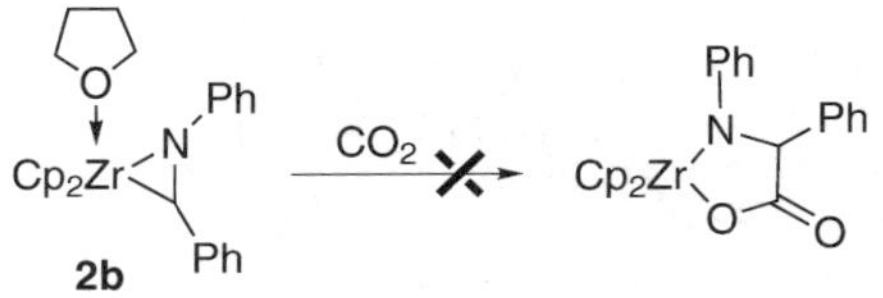

(23)

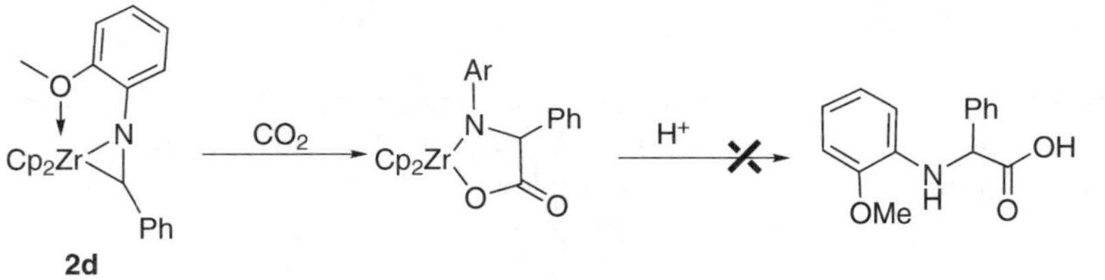

(24)

The insertion of isocyanates, isolelectronic with CO$_2$, proceeded more cleanly. Isocyanates (RNCO) with bulky R groups inserted exclusively into the Zr-C bond (Table 5); methanolysis resulted in α-amino amides **32** [21].

Other isocyanates inserted into the Zr-N bond as well as the Zr-C bond, resulting after workup in urea products **33** as well as the expected amides **32** [21]. The product ratio (Table 5) was independent of solvent and governed largely by the size of R. For R=i-Pr both products were formed but the major product was the amide. For small R (R=Et, Me, Bn) the major product was the urea.

Insertion into the Zr-N bond was even more common (compare lines 3 and 12 in Table 5) when the *N*-silyl zirconaaziridine **2a** was treated with RNCO, pre-

Synthesis and Reactivity of Zirconaaziridines

(25)

Table 5 RNCO insertion into zirconaaziridine **2b**

Entry	Zircona-aziridine	RNCO	Amide 32 (%)	Urea 33 (%)	Ratio 32/33
1	2b	t-Bu	76	0	100/0
2	2b	TMS	65	0	100/0
3	2b	i-Pr	69	8	90/10
4	2b	Et	31	52	37/63
5	2b	Me	23	45	34/66
6	2b	o-anisyl	24	31	44/56
7	2b	p-anisyl	28	48	37/63
8	2b	Bn	29	48	37/63
9	2b	Ph	20	49	29/71
10	2b	p-C$_6$H$_4$	20	53	27/73
11	2a	t-Bu	50	0	100/0
12	2a	i-Pr	0	56	0/100
13	2a	Ph	0	56	0/100
14	EBTHI analog of 2b	Ph	71	29	71/29
15	2d	Me	92	0	100/0

sumably because the effective steric bulk of the TMS substituent in **2a** is (due to the length of the N–Si bond) less than that of the Ph substituent in **2b**. The difference toward an entering electrophile is compounded by the fact that the two Cp rings in **2b** encourage the phenyl to remain in the Zr–N–C plane. Replacement of the Cp ligands with the more bulky rac-EBTHI ligand (compare entries 9 and 14 in Table 5) decreased insertion into the Zr–N bond, and the methoxy ligand in **2d** blocked it completely (albeit at a cost of substantial decrease in the rate of the insertion reaction) [21].

Carbodiimides, also isoelectronic with CO_2, react with zirconaaziridines exclusively through insertion into their Zr–C bonds. The resulting metallacycles

34 can be converted, after workup, into α-amino amidines – compounds which have recently attracted attention because of their potential biological activity and synthetic applications [49, 62, 63].

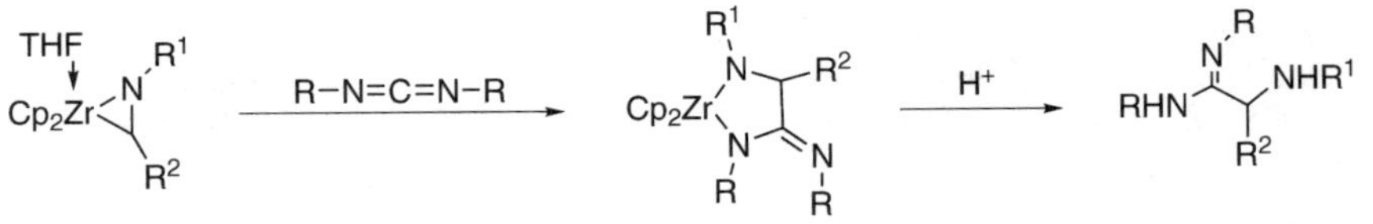

(26)

4.3.1
Mechanism of Heterocumulene Insertion; Implication for Other Electrophiles

Braunstein has suggested that precoordination is necessary for heterocumulene insertion [64]. If this suggestion is correct, THF must dissociate from **2a** or **2b**, or PMe$_3$ from **2o**, before insertion can take place. Such a mechanism is supported by the kinetics of the reaction (Eq. 27). The inverse dependence upon [THF] in the rate law (Eq. 28) reflects the reversible dissociation of THF from **2p** in Eq. 27 [49]. (The dissociation is an unfavorable, but rapidly maintained, preequilibrium.)

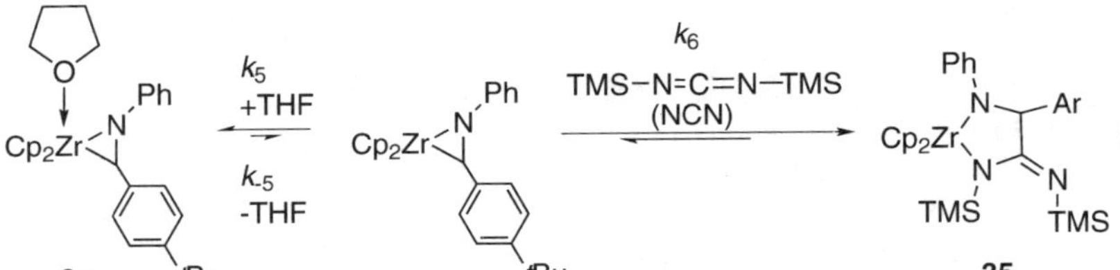

(27)

$$-\frac{d[\mathbf{2p}]}{dt} = \frac{K_5 k_6 [\text{NCN}][\mathbf{2p}]}{[\text{THF}]} \qquad (28)$$

where

$$K_5 = \frac{k_5}{k_{-5}} \qquad (28\text{a})$$

The reaction can be run in reverse at high [THF], with bis-(trimethylsilyl) carbodiimide replaced by THF. The rate is independent of [THF], consistent with rate-limiting extrusion of carbodiimide, i.e., with the operation of a similar dissociative mechanism in reverse. Similarly, addition of trimethylphosphine to **35** results in extrusion of the carbodiimide to form the phosphine-stabilized zirconaaziridine [49]. In contrast, the isocyanate insertions are effectively irreversible.

Isocyanate insertions presumably go through the same 16-electron intermediate, which explains why the Zr–N/Zr–C ratio from those reactions is independent of solvent. It is expected that the "inside" position (recall the discussion in Sect. 3.1) is favored thermodynamically for heterocumulene coordination (as it is for THF in **2a** and **2b** and PMe₃ in **2o**). Furthermore, inside coordination should lead to insertion in the Zr–N bond as in Eq. 29. The observed product ratios, which imply that Zr–N and Zr–C insertion are competitive, suggest that the inside complex **36-i** reacts more slowly than the outside complex **36-o**, i.e., that k_9 in Eq. 29 is $<k_{10}$. (It is also possible that insertion is faster than coordination, that **36-i/36-o** is kinetically controlled and equal to k_7/k_8, and that the kinetic ratio k_7/k_8 is closer to unity than the thermodynamic one.) Precoordination seems to be required for aldehydes and imines also.

(29)

4.4
Carbonates

Cyclic carbonates can be used as synthetic equivalents of CO_2. Their insertion into the Zr–C bond of zirconaaziridines yields spirocyclic organometallic complexes **37**. (Insertion into the Zr–N bond is never observed, perhaps because of the steric demands of the carbonate ring.) Methanolysis or protonolysis of the insertion product in THF yields amino acid ester **38**. *Methanolysis in benzene* results in cleavage of the organic fragment from the metal center as well as transesterification to the amino acid methyl ester **39**. Transesterification occurs only after treatment with methanol; other alcohols and water do not give analogous products, but result only in amino acid ester **38**. A crossover experiment confirmed **38** is an intermediate in the transesterification occurring during methanolysis [21].

5
Control of Stereochemistry Resulting from Insertion Reactions

The synthesis of amino acid esters can be carried out enantioselectively when optically active EBTHI zirconaaziridines are used. After diastereomeric zirconaaziridines are generated and allowed to equilibrate (recall Scheme 3), the stereochemistry of the chiral carbon center in the insertion product is determined by competition between the rate constants k_{SSR} and k_{SSS} for the epimerization of zirconaaziridine diastereomers and the rate constants $k_R[EC]$ and $k_S[EC]$ for ethylene carbonate (EC) insertion (Eq. 31) [43]. When $k_R[EC]$ and $k_S[EC]$ are much greater than k_{SSR} and k_{SSS}, the product ratio reflects the equilibrium ratio as shown in Eq. 32. However, the opposite limit, where epimerization is much faster than insertion, is a Curtin–Hammett kinetic situation [65] where the product ratio is given by Eq. 33.

$$\frac{SSS-18}{SSR-18} = K_{eq} = \frac{k_{SSR}}{k_{SSS}} = \frac{SSR-17}{SSS-17} \tag{32}$$

$$\frac{SSS-18}{SSR-18} = K_{eq}\frac{k_S}{k_R} \tag{33}$$

In Fig. 11, at high concentrations of ethylene carbonate, the rate constants $k_S[EC]$ and $k_R[EC]$ for insertion into the EBTHI zirconaaziridine **17q** are much greater than k_{SSR} and k_{SSS} and insertion occurs more rapidly than the equilibrium can be maintained. The product ratio reflects the equilibrium of **17q**, where K_{eq} is 17.2 (Eq. 32) [21]. Beak has called this limit a "dynamic thermodynamic resolution" pathway [66]. In contrast, at the lowest concentration of ethylene carbonate in Fig. 10, the first-order rate constants k_{SSR} and k_{SSS} for diastereomer interconversion are comparable to the effective first-order rate constants for insertion. As K_{eq} is known to be 17.2, k_S/k_R can be calculated; the 53% ee of (S)-amino acid ester **19q** (Scheme 9) implies that $k_S/k_R<0.19$ (Eq. 33) and that the rate constant for insertion $k_R[EC]$ into the minor diastereomer is at least five times faster than $k_S[EC]$ into the major diastereomer.

Similarly, a 92% ee of (S)-amino acid ester **19r** is obtained after the equilibrated zirconaaziridine **17r** is treated with high concentrations of ethylene carbonate; K_{eq} is approximately 24.0. The reaction of equilibrated **17r** with low concentrations of ethylene carbonate yields the (S)-amino acid ester **19r** in 98% ee, implying that $k_S/k_R<4.1$ in this case and the rate constant of insertion $k_S[EC]$ into the major diastereomer is at least four times faster than its counterpart $k_R[EC]$ into the minor diastereomer.

The stereochemistry of the chiral center in the insertion product can also be controlled by using an optically active, C_2-symmetric, cyclic carbonate (e.g.,

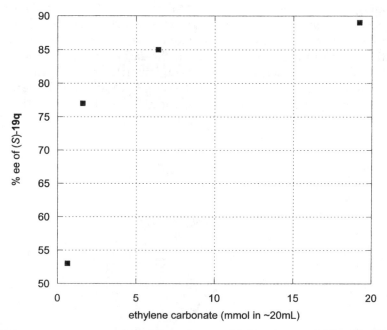

Fig. 11 Percent ee vs amount of ethylene carbonate added to zirconaaziridine **17q**. Zirconaaziridine was prepared at 70 °C and treated with ethylene carbonate at room temperature

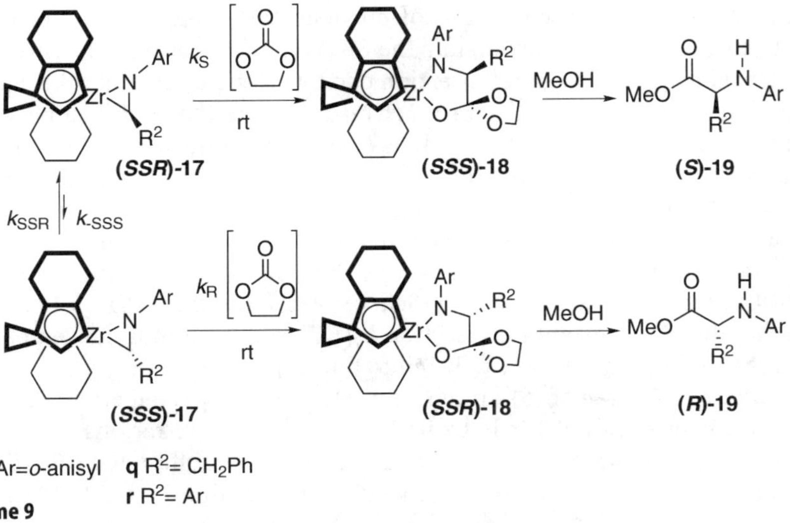

Scheme 9

Ar=o-anisyl
q R^2 = CH_2Ph
r R^2 = Ar

R,R-diphenyl ethylene carbonate (R,R-DPEC)) with a racemic zirconaaziridine. (C_2-symmetric, cyclic carbonates are attractive as optically active synthons for CO_2 because optically active diols are readily available through Sharpless asymmetric dihydroxylations [67].) Reaction through diastereomeric transition states affords the two diastereomers of the spirocyclic insertion product; protonolysis and Zr-mediated transesterification in methanol yield α-amino acid esters. As above, the stereochemistry of the new chiral center is determined by the competition between the rate of interconversion of the zirconaaziridine enantiomers and the rate of insertion of the carbonate. As the ratio of zirconaaziridine enantiomers (S)-2/(R)-2 is initially 1:1, a kinetic quench of their equilibrium will result in no selectivity (see Eq. 32). Maximum diastereoselectivity (and, therefore, maximum enantioselectivity for the preparation of the

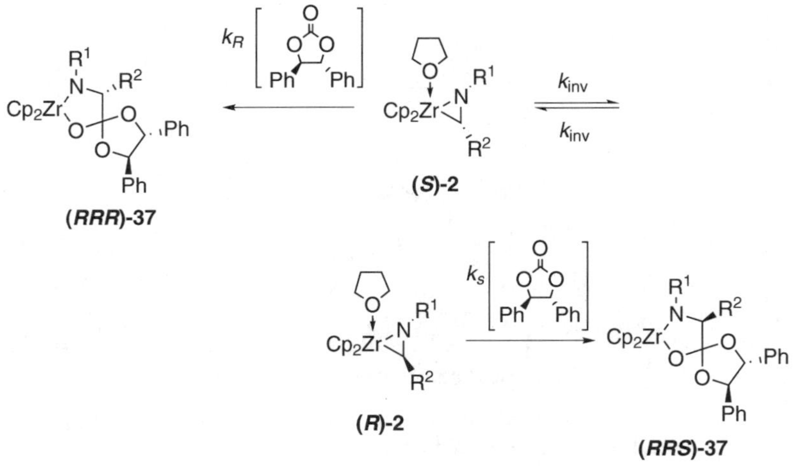

(34)

amino acid esters) will be observed when insertion is slow relative to the rate of enantiomer interconversion. Under these conditions the selectivity (see Eq. 33) will be k_R/k_S, also known as the selectivity factor s.

The factor s can be determined by treating racemic **2** with racemic carbonate (Hoffman test) (Scheme 10, Eq. 35) [68].

$$\frac{k_{\text{fast}}}{k_{\text{slow}}} = s = \frac{SSS + RRR}{RSS + SRR} \tag{35}$$

Obtaining the maximum diastereomer ratio s often requires slow addition, with a syringe pump, of the optically active carbonate to the zirconaaziridine. When run under these conditions (Table 6), the reaction can be termed a "dynamic

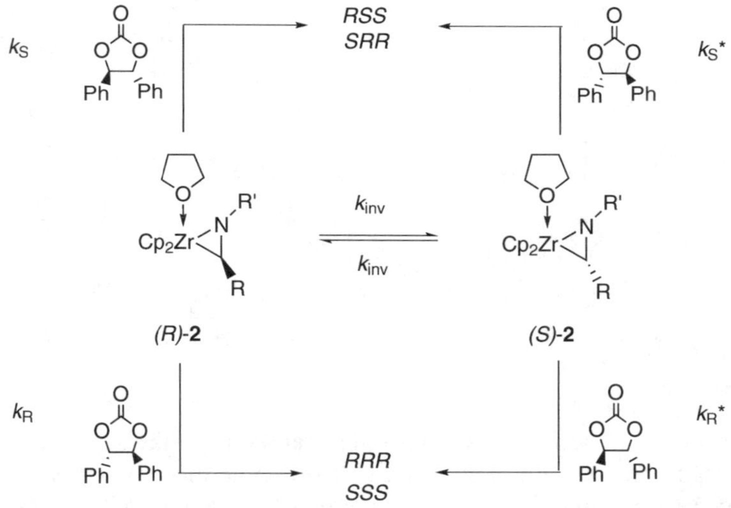

Scheme 10 *The rate constants k_R and k_S are defined for (R,R)-DPEC insertion into the Zr–C bond of zirconaaziridines **2**, but also apply to the enantiomeric transition states

Table 6 Predicted and observed de values for zirconaaziridine reaction with (RR)-DPEC (* Maximum de predicted by Hoffman test)

Complex	R¹	R²	Maximum de (%) predicted*	Observed de (%)
2a	TMS	Ph	90	90
2b	Ph	Ph	76	76
40a	TMS	CH$_2$Ph	82	77
40b	Ph	CH$_2$Ph	74	71
40c	TMS	CH$_2$CH(CH$_3$)$_2$	21	18

kinetic resolution". However, strictly speaking, this term should be applied only to situations where the starting material becomes resolved. More applicable to the present situation is the term "dynamic kinetic asymmetric transformation", coined by Trost [69], where the starting material has been transformed into another compound in an enantioselective fashion.

Silylated amino acid esters have been prepared by a dynamic kinetic asymmetric transformation of this type. Slow addition (by syringe pump) of (*R*,*R*)-DPEC maintained insertion slow relative to interconversion and afforded the silyl-substituted amino acid esters in Eq. 36 with ee values of 80, 68, and 83%. The ester **38c** (R=TMS) was too base-sensitive to permit removal of the chiral diol, but transesterification of the other β-hydroxyethyl esters **38** to the corresponding methyl esters was straightforward [22].

$$(36)$$

2c R = TMS
2s R = CH$_2$TMS
2t R = *p*-C$_6$H$_4$TMS

37c, 38c R = TMS   80% de
37s, 38s R = CH$_2$TMS   68% de
37t, 38t R = *p*-C$_6$H$_4$TMS   83% de

It is instructive to perform a similar kinetic analysis on the Taguchi system in Sect. 2.4. In view of the Curtin–Hammett principle the product ratio (*RRR*)-**39**:(*SSR*)-**39** need not reflect the diastereomer ratio (*SR*)-**2n**:(*RR*)-**2n** (as Taguchi and coworkers assumed it did) [31]. However, when **2n** was generated at 67 °C, both benzaldehyde and acetaldehyde gave the same ratio (5:95, *RRR:SSR*) of the diastereomers of the amino alcohols **39**, suggesting that the assumption made by Taguchi and coworkers is correct, that Eq. 32 applies, and that *RR*-**2n** and *SR*-**2n**

39	T(°C)	dr, RRR:SSR
a R=Ph	a 23	88:12
b R=Me	a 67	5:95
	b 23	87:13
	b 67	5:95

$$(37)$$

are at equilibrium with K_{eq}=19. When **2n** was generated at 23 °C, both benzaldehyde and acetaldehyde gave *RRR*-**39**:*SSR*-**39** ratios of 87:13. The major diastereomer at 23 °C was thus the opposite of that observed at 67 °C – suggesting that (*SR*)-**2n**:(*RR*)-**2n** is 87:13 at 23 °C but that these two diastereomers are not in equilibrium (the more stable diastereomer of **2n** at 67 °C is surely also the more stable diastereomer at 23 °C).

6
Interconversion of Stereochemistry at Zirconaaziridine Carbons

It would be preferable to carry out a dynamic kinetic asymmetric transformation on a zirconaaziridine without having to add the inserting reagent slowly. The fact that the enantiomers of **2i** interconvert rapidly makes it possible to carry out such a transformation at maximum de (64% with *R,R*-DPEC) without having to use a syringe pump for addition of the optically active carbonate.

This fast enantiomer interconversion probably occurs because of the facile isomerization of **2i** to the corresponding azaallyl hydride [26]. Such isomerizations are common when the carbon of a zirconaaziridine bears a primary or secondary alkyl substituent. Generally the equilibrium favors the azaallyl hydride, but the room-temperature NMR spectrum of **2i** shows only the zirconaaziridine; apparently the steric congestion around the carbon in **2i** changes the position of the equilibrium. Even an azaallyl hydride that dominates the equilibrium, as **40** does (Fig. 12), appears to react with electrophiles through the corresponding zirconaaziridine. (There may also be a direct associative path when **40** reacts with carbonates.) [24].

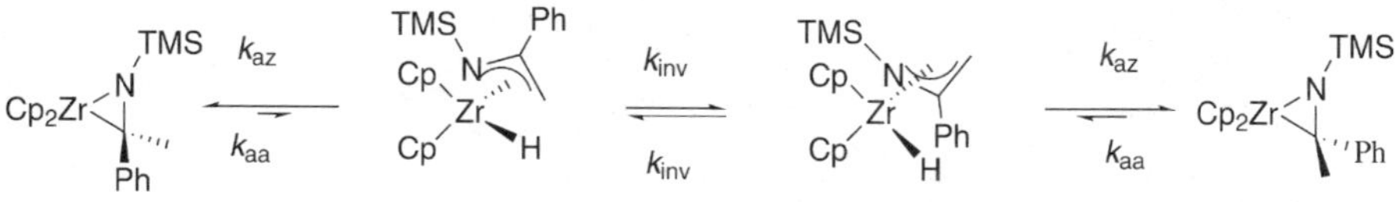

$$(38)$$

The kinetics of the reaction of **40b** with PMe$_3$ (Eq. 39) have been examined. The rate does not vary with [PMe$_3$] from 0.36 to 0.73 M, showing that the reaction is not associative in PMe$_3$ and suggesting the high PMe$_3$ limit (Eq. 41) of a dissociative rate law (Eq. 40). In this limit k_{obs}=k_{az}, the rate constant for for-

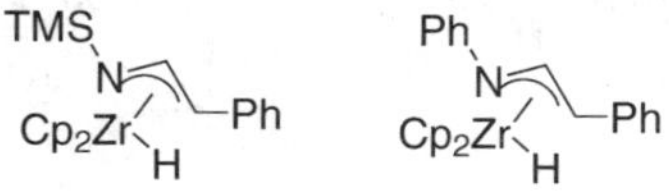

Fig. 12 Azaallyl hydrides

mation of a ligand-free zirconaaziridine from the azaallyl hydride **40b**. Thus, k_{az} at 25 °C is 6.1(1)×10^{-3} s^{-1} with a $\Delta H^{\ddagger}$ of +23.2(1) kcal/mol and a $\Delta S^{\ddagger}$ of +9.2(3) eu. As the corresponding zirconaaziridine is not visible in the ^{1}H NMR spectrum of **40b**, one can estimate that $k_{aa}/k_{az}>20$ and thus that $k_{aa}>0.12$ s^{-1} [24], as summarized in Scheme 11.

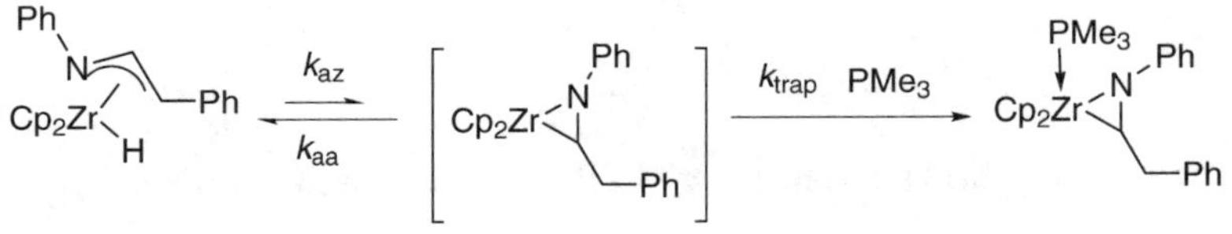

$$\frac{-d[\mathbf{40b}]}{dt} = \frac{k_{az}k_{trap}[\text{PMe}_3][\mathbf{40b}]}{k_{aa} + k_{trap}[\text{PMe}_3]} \tag{40}$$

$$k_{trap}[\text{PMe}_3] \gg k_{aa} : \frac{-d[\mathbf{40b}]}{dt} = k_{az}[\mathbf{40b}] \tag{41}$$

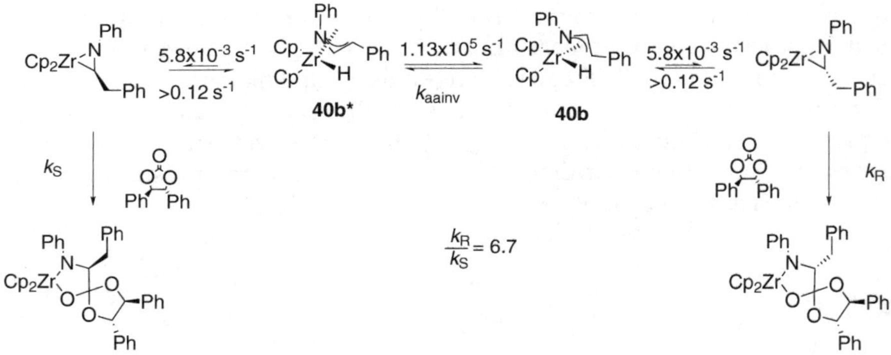

Scheme 11

Interconversion of enantiomeric azaallyl hydrides like **40** and **40*** is facile and probably involves an isomerization from an η^3 to an N-bound η^1 structure with a plane of symmetry (Eq. 42) [24]. This process is analogous to the well-documented π-σ-π allyl isomerization [70].

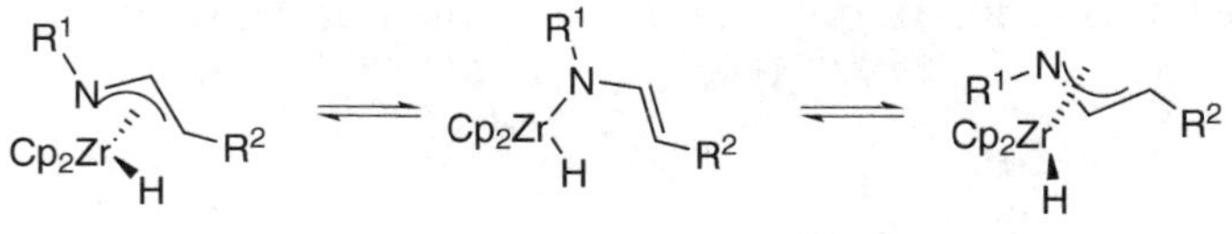

Rate constants for inversion, k_{Cp}, have been measured (Table 7) for **2i, 40a**, and **40b** by observing the broadening and coalescence of the Cp signals[24]. For the

Synthesis and Reactivity of Zirconaaziridines

Table 7 Activation parameters and kinetic data for azaallyl hydride inversion

Complex	$\Delta H^{\ddagger}$ kcal/mol	$\Delta S^{\ddagger}$ eu	$\Delta G^{\ddagger}$ kcal/mol	k_{Cp}, s^{-1}, 298 K
2i	10.0(6)	–20(2)	4.0	10
40a	10.2(1)	1.9(8)	9.6	5.6(7)×10^5
40b	12.9(1)	7.9(4)	10.3	1.1(1)×10^5

azaallyl hydrides **40a** and **40b**, k_{Cp} is the rate constant k_{aainv} for azaallyl hydride interconversion, but for **2i** – where the stable form is the zirconaaziridine – k_{Cp} reflects the barrier to isomerization and inversion combined.

The interconversion of zirconaaziridine enantiomers is slower when there is not a primary or secondary alkyl on the zirconaaziridine carbon and isomerization to an azaallyl hydride is not possible. A mechanism that remains available involves the isomerization of each enantiomer to a planar η^1 complex (Eq. 43), such as that known to interconvert the enantiomers of aromatic aldehydes [16]. For the chelated zirconaaziridine **2d**, high-level density functional theory (DFT) methods and a continuum solvation model have shown that enantiomer interconversion occurs through an η^1-imine intermediate (**A**) rather than through homolysis (**B**) or heterolysis (**C**) of the Zr–C bond [71] (Fig. 13).

$$\begin{array}{c}\text{Cp}_2\text{Zr structures} \end{array} \quad (43)$$

(*R*)-2   (*S*)-2

It is possible to determine the rate constant k_{zainv} for zirconaaziridine enantiomer interconversion directly by examining an unequal mixture of zirconaaziridine enantiomers. Treating a racemic zirconaaziridine with half an equivalent of the (*R,R*) carbonate will (unless k_{zainv} is *faster* than insertion) generate such an unequal mixture in the remaining half equivalent of the zirconaaziridine (Scheme 12). Thus racemic **2b**, treated with half an equivalent of (*R,R*)-DPEC (known to give primarily the (*RRS*) azazirconacycle) will leave primarily (*R*)-**2b**, and the racemization of the latter can be directly observed. The decay of the optical activity (which will occur with a rate constant of 2 k_{zainv}) can be moni-

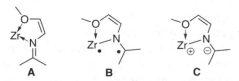

Fig. 13 Possible intermediates in the mechanism of zirconaaziridine interconversion

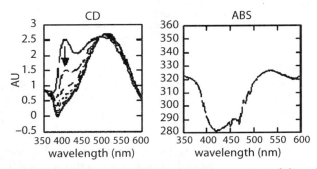

Fig. 14 The racemization and, therefore, enantiomer interconversion of **2b** can be monitored directly by CD

tored by circular dichroism (CD), as Fig. 14 demonstrates. (The time-invariant CD spectrum at longer wavelengths belongs to the insertion product.) In the experiment shown insertion was complete before the first CD was taken, so the electronic absorption spectrum remained unchanged during the decay of the CD signal (about 20 min) [72].

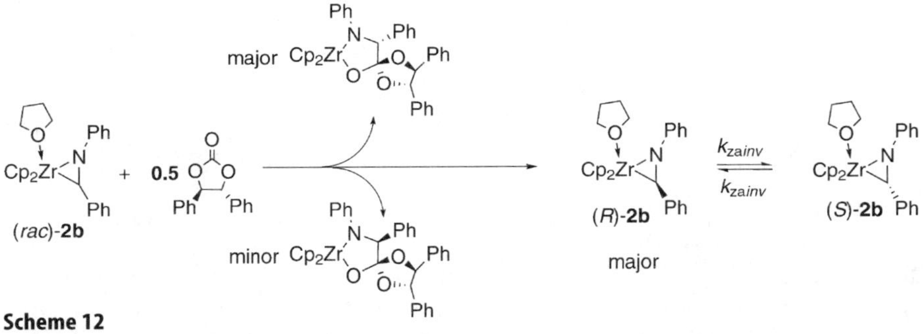

Scheme 12

7
Thermal Transformations of Zirconocene-η^3-1-Azaallyl Hydrides

Azaallyl hydrides undergo many reactions as well as enantiomer interconversion. Heating **40a** at 70 °C overnight led to the C–H activation product **41**, identified by ^{1}H and ^{13}C NMR (Eq. 44) [24]. Similar C–H activations (Eq. 45) have been observed by Andersen [73] and Scott [74].

Monitoring the progress of the reaction in Eq. 44 by ^{1}H NMR revealed the formation of a transient species, assigned structure **42a** on the basis of its ^{1}H and ^{13}C NMR spectra. The ^{13}C NMR spectrum of **42a** shows a resonance at 184.8 ppm, consistent with formation of a Zr–phenyl bond. Complete conversion to a related product **42b** was obtained with the azaallyl hydride **40b**

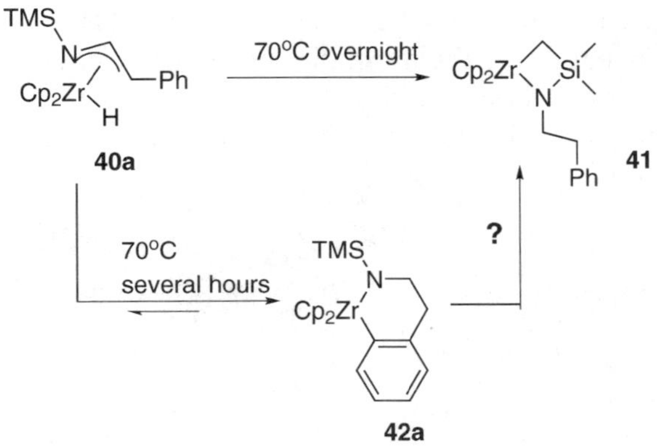

Synthesis and Reactivity of Zirconaaziridines

(44)

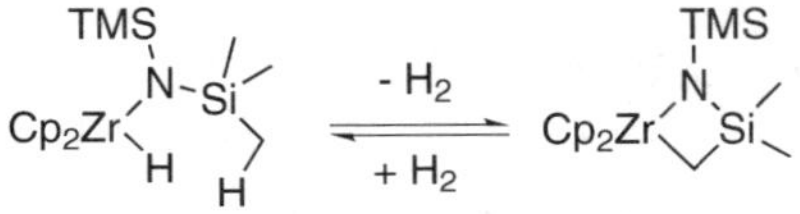

(45)

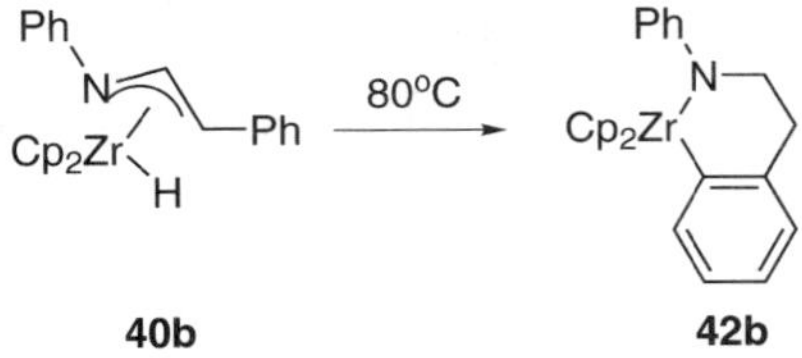

(46)

(Eq. 46). Both **42a** and **42b** show downfield Zr–C resonances around 182 ppm in the ^{13}C NMR spectrum [24].

H$_2$ evolution is not observed, so these products probably result from the competition between oxidative addition of different C–H bonds to the Zr(II) center (Scheme 13). The fact that **41** is thermodynamically more stable than **42a** probably results from the ability of a silyl substituent to stabilize negative charge [75].

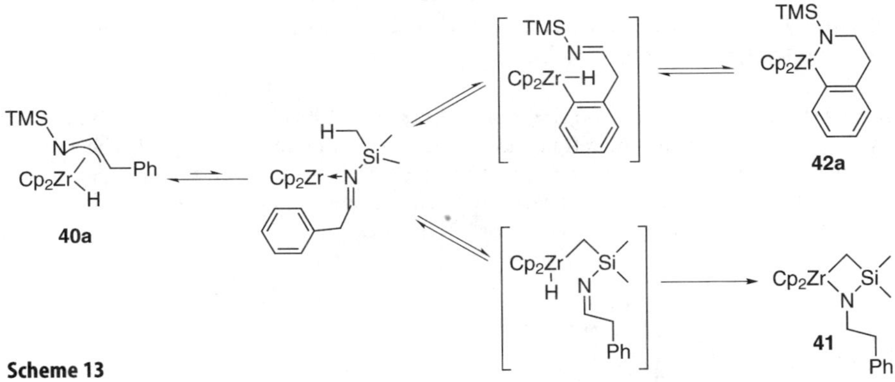

Scheme 13

Similarly, attempts to generate the azaallyl complex **40l** from *N*-TMS phenyl propyl amine result in formation of the aromatic metallation product **43** after only 45 min at 65 °C (Scheme 14). Interception of the intermediate **2l** with (*rac*)-DPEC gives the homophenylalanine ester **38l** in 96% de and 53% yield [24].

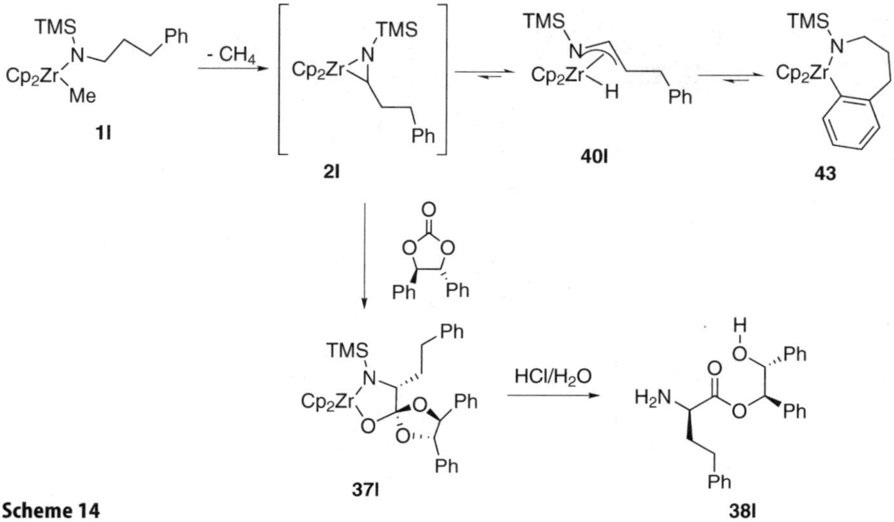

Scheme 14

8
Conclusions

Zirconaaziridines, which can be prepared in a few steps from commercially available starting materials, have polar Zr–C bonds that permit the insertion of many electrophilic reagents (C=O, C=N) as well as alkynes (C≡C) and alkenes (C=C). Appropriate substituents on their ring carbons make zirconaaziridines chiral and raise the possibility of their use in asymmetric synthesis. The lability of their chiral centers distinguishes zirconaaziridines from other organometallic reagents and makes it possible for them to undergo "dynamic kinetic asymmetric transformations".

Acknowledgements Research on zirconaaziridines in the Norton group has been supported by the National Science Foundation, most recently under grant CHE-02-1131. The authors are also grateful to Professor A.G. Barrett for drawing their attention to amino amidines, to Dr. Daniel A. Gately for early work on zirconaaziridines, and to Dr. Masanori Iimura for providing assistance in the preparation of this manuscript.

References

1. Ti: (a) Gao Y, Yoshida Y, Sato F (1997) Synlett 1353; (b) Fukuhara K, Okamoto S, Sato F (2003) Org Lett 5:2145; (c) Durfee LD, Hill JE, Fanwick PE, Rothwell IP (1990) Organometallics 9:75; (d) Durfee LD, Fanwick PE, Rothwell IP, Folting K, Huffman JC (1987) J Am Chem Soc 109:4720
2. Hf: Scott MJ, Lippard SJ (1997) Organometallics 16:5857
3. V: Sielisch T, Behrens U (1986) J Organomet Chem 310:179
4. Nb: (a) Roskamp EJ, Pedersen SF (1987) J Am Chem Soc 109:6551; (b) Roskamp EJ, Pedersen SF (1987) J Am Chem Soc 109:3152
5. Ta: (a) Mayer JM, Curtis CJ, Bercaw JE (1983) J Am Chem Soc 105:2651; (b) Takahashi Y, Onoyama N, Ishikawa Y, Motojima S, Sugiyama K (1978) Chem Lett 525
6. Mo: (a) Cameron TM, Ortiz CG, Abboud KA, Boncella JM, Baker RT, Scott BL (2000) J Chem Soc Chem Commun 7:573; (b) Okuda J, Herberich GE, Raabe E, Bernal I (1988) J Organomet Chem 353:65
7. W: Chiu KW, Jones RA, Wilkinson G, Galas AMR, Hursthouse MB (1981) J Chem Soc Dalton Trans 2088
8. Ru: Polm LH, van Koten G, Elsevier CJ, Vrieze K, van Santen BF, Stam CH (1986) J Organomet Chem 304:353
9. Ni: Hoberg H, Gotz, V, Kruger C, Tsay YH (1979) J Organomet Chem 169:209
10. Pt: Browning J, Empsall HD, Green M, Stone FGA (1973) J Chem Soc Dalton Trans 381
11. Sm: (a) Hou Z, Yoda C, Koizumi T, Nishiura M, Wakatsuki Y, Fukuzawa S, Takats J (2003) Organometallics 22:3586; (b) Imamoto T, Nishimura, S (1990) Chem Lett 203
12. Yb: (a) Takaki K, Tanaka S, Fujiwara Y (1991) Chem Lett 493; (b) Makioka Y, Taniguchi Y, Kitamura T, Fujiwara Y, Saiki A, Takaki K (1996) Organometallics 15:5476; (c) Makioka Y, Taniguchi Y, Kitamura T, Fujiwara Y, Saiki A, Takaki K (1997) Bull Chim Soc Fr 134:349; (d) Takaki K, Takeda M, Koshoji G, Shishido T, Takehira K (2001) Tetrahedron Lett 42:6357; (e) Takaki K, Koshoji G, Komeyama K, Takeda M, Shishido T, Kitani A, Takehira K (2003) J Org Chem 68:6554
13. For a review see: Seebach D (1979) Angew Chem Int Ed Engl 18:239
14. Erker G, Dorf D, Czisch P, Petersen JL (1986) Organometallics 5:668
15. Waymouth RM, Clauser KR, Grubbs RH (1986) J Am Chem Soc 108:6385
16. Gladysz JA, Boone BJ (1997) Angew Chem Int Ed Engl 36:550
17. Wailes PC, Weigold H, Bell AP (1971) J Organomet Chem 33:181
18. Grossman RB, Davis WM, Buchwald SL (1991) J Am Chem Soc 113:2321
19. For inversion barriers in silylamines, see: (a) Gordon MS (1986) Chem Phys Lett 126:451; (b) Albert K, Roesch N (1997) Chem Ber Recl 130:1745
20. Buchwald SL, Wanamaker MW, Watson BT, Dewan JC (1989) J Am Chem Soc 111:4486
21. Gately DA, Norton JR, Goodson, PA (1995) J Am Chem Soc 117:986
22. Chen J-X, Tunge JA, Norton JR (2002) J Org Chem 67:4366
23. Coles N, Harris MCJ, Whitby RJ, Blagg J (1994) Organometallics 13:190
24. Tunge JA (2000) PhD thesis, Columbia University
25. Buchwald SL, Nielsen RB (1988) J Am Chem Soc 110:3171
26. Harris MCJ, Whitby RJ, Blagg J (1994) Tetrahedron Lett 35:2431
27. Negishi E, Cederbaum FE, Takahashi T (1986) Tetrahedron Lett 27:2829
28. Jensen M, Livinghouse T (1989) J Am Chem Soc 111:4495
29. Soleil F, Choukroun R (1997) J Am Chem Soc 119:2938
30. Dioumaev VK, Harrod JF (1997) Organometallics 16:1452
31. Ito H, Taguchi T, Hanzawa Y (1992) Tetrahedron Lett 33:4469
32. Lubben TV, Plössl K, Norton JR, Miller MM, Anderson OP (1992) Organometallics 11:122

33. Kempe R, Spannenberg A, Lefeber C, Zippel T, Rosenthal U (1998) Z Kristallogr New Crystal Structures 213:791
34. (a) Rosenthal U, Ohff A, Michalik M, Görls H, Burlakov VV, Shur VB (1993) Angew Chem Int Ed Engl 32:1193; (b) Nitschke JR, Zurcher S, Tilley TD (2000) J Am Chem Soc 122:10345
35. Wolczanski PT, Bercaw JE (1979) J Am Chem Soc 101:6450
36. Davis JM, Whitby RJ, Jaxa-Chamiec A (1992) Tetrahedron Lett 33:5655
37. Davis JM, Whitby RJ, Jaxa-Chamiec A (1994) Tetrahedron Lett 35:1445
38. (a) Tilley TD (1985) Organometallics 4:1452; (b) Tilley TD (1985) J Am Chem Soc 107:4084; (c) Elsner FH, Woo HG, Tilley TD (1988) J Am Chem Soc 110:313; (d) Campion BK, Falk J, Tilley TD (1987) J Am Chem Soc 109:2049; (e) Elsner FH, Tilley TD, Rheingold AL, Geib SJ (1988) J Organomet Chem 358:169; (f) Heyn RH, Tilley TD (1989) Inorg Chem 28:1768; (g) Cardin DJ, Keppie SA, Kingston BM, Lappert MF (1967) J Chem Soc Chem Commun 1035; (h) Xue Z, Li L, Hoyt LK, Diminnie B, Pollitte JL (1994) J Am Chem Soc 116:2169; (i) Procopio LJ, Carroll PJ, Berry DH (1991) J Am Chem Soc 116:2169; (j) Procopio LJ, Carroll PJ, Berry DH (1993) Organometallics 12:3087
39. Honda T, Satoh S, Mori M (1995) Organometallics 14:1548
40. Honda T, Mori M (1996) Organometallics 15:5464
41. Honda T, Mori M (1996) J Org Chem 61:1196
42. Grossman RB (1991) PhD thesis, Massachusetts Institute of Technology
43. Gately DA, Norton JR (1996) J Am Chem Soc 118:3479
44. Gately DA, Norton JR, Goodson PA (unpublished results)
45. Tatsumi K, Nakamura A, Hofmann P, Stauffert P, Hoffman R (1985) J Am Chem Soc 107:4440
46. Allen FH, Kennard O, Watson DG, Brammer L, Orpem AG, Taylor R (1987) J Chem Soc Perkin Trans II S1
47. Atwood JL (1983) Organometallics 2:770
48. Cadierno V, Zablocka M, Donnadieu B, Igau A, Majoral J-P, Skowronska A (1999) J Am Chem Soc 121:11086
49. Tunge JA, Czerwinski CJ, Gately DA, Norton JR (2001) Organometallics 20:254
50. Tunge JA, Gately DA, Norton JR (1999) J Am Chem Soc 121:4520
51. (a) Hart DW, Blackburn TF, Schwartz J (1975) J Am Chem Soc 97:679; (b) Nugent WA, Calabrese JC (1984) J Am Chem Soc 106:6422; (c) Nugent WA, Thorn DL, Harlow RL (1987) J Am Chem Soc 109:2788
52. Coles N, Whitby RJ, Blagg J (1990) Synlett 271
53. Coles N, Whitby RJ, Blagg J (1992) Synlett 143
54. Harris MCJ, Whitby RJ, Blagg J (1995) Tetrahedron Lett 36:4287
55. Barluenga J, Sanz R, Fañanás FJ (1997) J Org Chem 62:5953
56. Buchwald SL, Wanamaker MW, Watson BT (1989) J Am Chem Soc 111:776
57. Knorr L (1884) Chem Ber 17:1635 and Paal C (1885) Chem Ber 18:367
58. Dhawan R, Arndtsen BA (2004) J Am Chem Soc 126:468
59. (a) Takahashi T, Li Y (2002) Zirconacyclopentadienes in organic synthesis. In: Marek I (ed) Titanium and zirconium in organic synthesis. Wiley-VCH, Weinheim, Sect 2.3.1; (b) Venazi LM, Lehmann R, Keil R, Lipshutz BH (1992) Tetrahedron Lett 33:5857; (c) Takahashi T, Kotora M, Kasai K, Suzuki N (1994) Tetrahedron Lett 35:5685; (d) Kotora M, Kasai K, Suzuki N, Takahashi T (1995) J Chem Soc Chem Commun 109
60. Dekura F, Honda T, Mori M (1997) Chem Lett 8:825
61. Buchwald SL (personal communication)
62. Keung W, Bakir F, Patron AP, Rogers D, Priest CD, Darmohusodo V (2004) Tetrahedron Lett 45:733
63. Katrizky AR, Button MA, Busont S (2001) J Org Chem 66:2865

64. Braunstein P, Nobel D (1989) Chem Rev 89:1927
65. (a) Seeman JI (1983) Chem Rev 83:83; (b) Eliel EL, Wilen SH (1994) Stereochemistry of organic compounds. Wiley, New York, p 648; (c) Carey FA, Sundberg RJ (2000) Advanced organic chemistry, 3rd edn. Plenum, New York, p 220
66. Beak P, Anderson DR, Curtis MD, Laumer JM, Pippel DJ, Weisenburger GA (2000) Acc Chem Res 33:715
67. Wang ZM, Sharpless KB (1994) J Org Chem 59:8302
68. Hirsh R, Hoffmann RW (1992) Chem Ber 125:975
69. Trost BM, Toste FD (1999) J Am Chem Soc 121:3543
70. Bercaw JE, Moss JR (1992) Organometallics 11:639
71. Baik M-H, Frost BJ, Friesner RA, Norton JR (manuscript in preparation)
72. Cummings SA, Iimura M, Norton JR (unpublished results)
73. (a) Planalp RP, Andersen RA (1983) Organometallics 2:1675; (b) Planalp RP, Andersen RA, Zalkin A (1983) Organometallics 2:16; (c) Simpson SJ, Andersen RA (1981) Inorg Chem 20:3627
74. Morton C, Munslow IJ, Sanders CJ, Alcock NW, Scott P (1999) Organometallics 18:4608
75. (a) Brinkman EA, Berger S, Brauman JI (1994) J Am Chem Soc 116:8304; (b) Lambert JB (1990) Tetrahedron 46:2677

Synthesis and Reactivity of Zirconium–Silene Complex

Miwako Mori (✉)

Graduate School of Pharmaceutical Sciences, Hokkaido University, Sapporo 060-0812, Japan
mori@pharm.hokudai.ac.jp

1	Introduction	42
1.1	Reaction of Alkyne with Cp_2ZrCl_2 and $Ph_2{}^tBuSiLi$	45
2	Formation of Zirconium–Silene Complex	48
2.1	Reaction of Alkyne with Cp_2ZrCl_2 and $Me_2PhSiLi$	48
2.2	Possible Reaction Course for the Formation of Zirconium–Silene Complex	51
2.3	Confirmation of the Reaction Mechanism	53
2.4	Reactivity of Silazirconacyclopentene Formed from a Zirconium–Silene Complex	56
2.4.1	Insertion of Isocyanide into Silazirconacyclopentene	56
2.4.2	Insertion of Carbon Monoxide	58
2.4.3	Transmetalation of Zirconium to Copper of Silazirconacyclopentene	59
3	Perspective	60
	References	61

Abstract When Cp_2ZrCl_2 was treated with 2 equivalents of $Me_2PhSiLi$, a zirconium–silene complex was formed. In the presence of alkyne, diarylalkyne reacted with silene coordinated with zirconium to give silazirconacyclopentene, which was hydrolyzed to give vinylsilane. On the other hand, dialkylalkyne inserted into zirconium–silene complex to give silazirconacyclopentene, which was hydrolyzed to give allylsilane. Formation of the zirconium–silene complex was monitored by ^1H NMR spectra.

Keywords Zirconium–silene complex · Silene · Azazirconacyclopropane · Azazirconacyclopentene · $Cp_2Zr(SiMe_2Ph)$

Abbreviations
Ad   Adamantyl
Ar   Aryl
Bu   Butyl
^tBu tert-Butyl
°C   Degrees Celsius
Cp   Cyclopentadienyl
Cp*   Pentamethylcyclopentadienyl
Cy   Cyclohexyl
Et   Ethyl
h   Hour(s)
Ln   Ligand

Me Methyl
MS Mass spectrometry
NMR Nuclear magnetic resonance
Ph Phenyl
Pr Propyl
iPr Isopropyl
rt Room temperature
Temp. Temperature
THF Tetrahydrofuran
TMS Trimethylsilyl

1
Introduction

Silene, which has a silicon–carbon double bond, is a very reactive organosilicon species, and isolation of silene is very difficult [1]. The first report showing the presence of silene as an intermediate was the formation of 1,3-disilacyclobutane and ethylene resulting from pyrolysis of 1,1-disubstituted-1-silacyclobutanes by Gusel'nikov and Flowers. They reported that an unstable compound containing a silicon–carbon double bond should be formed and that dimerization of this species resulted in the formation of 1,3-disilacyclobutane (Eq. 1) [2]. Hydrolysis of intermediary silene afforded silanol. Gusel'nikov also confirmed the formation of silacyclohexene by Diels–Alder reaction of intermediary silene and diene (Eq. 2) [3]. In 1981, Brook first succeeded in the isolation of a thermodynamically stable silene, 2-adamantyl-2-trimethylsiloxy-1,1-bis(trimethylsilyl)-1-silaethene, which has a carbon–silicon double bond, by photolysis of a solution of acyl silane. This silene has a sterically bulky adamantyl group on the carbon of the silicon–carbon double bond and its structure was confirmed by X-ray diffraction analysis at –50 °C (Eq. 3) [4]. Methanol and water readily add across the silicon–carbon double bond to give an addition product. Wiberg obtained a stable silene by thermal salt elimination of LiF (Eq. 4) [5]. In this case, the bulky silyl group was placed on the carbon of silene. In a similar manner, Apeloig synthesized stable silenes by the reaction of silyl lithium derivatives and adamantanone (Eq. 5) [6]. This silene reacted with 1-methoxybutadiene to produce the expected Diels–Alder product. Recently, theoretical studies on silene have been carried out [7].

[Equation (2): silacyclobutane → silene intermediate + ethylene/isoprene → silacyclohexene products]

[Equation (3): (Me₃Si)₃Si-C(O)Ad ⇌ (hν) Me₃Si-Si(OSiMe₃)=C(SiMe₃)Ad, then ROH → Me₃Si-Si(OSiMe₃)(RO)-C(SiMe₃)(H)Ad, Ad=adamantyl]

[Equation (4): Me-Si(Me)(SiMe^tBu₂)(F)(Li)-SiMe₃ —ca. 100 °C→ Me₂Si=C(SiMe^tBu₂)(SiMe₃)]

[Equation (5): R'-Si(SiR₃)(R'')-Li·3THF + adamantanone → Bu^tMe₂Si-Si(RMe₂Si)-adamantylidene; with CH₂=CH-CH=CH-OMe → ^tBuMe₂Si-Si(Me₂RSi)(adamantyl)-cyclohexenyl-OMe]

Since silene is an unstable species, various transition metal–silene complexes coordinated by the silicon–carbon double bond have been reported. In 1970, Pannel reported the formation of silene by irradiation of an iron complex (Eq. 6) [8]. He obtained an iron–TMS complex that was apparently formed from silene and an iron–hydride complex generated from the starting iron complex by β-hydrogen elimination [8]. Wrighton confirmed the existence of tungsten- and iron–silene complexes by examination of NMR spectra obtained at low temperature (Eqs. 7 and 8) [9].

$$CpFe(CO)_2CH_2SiMe_2H \xrightarrow[UV]{PPh_3} CpFe(CO)PPh_3SiMe_3 \quad (6)$$

$$[=Si\overset{Me}{\underset{Me}{}} + FeH]$$

$$CpW(CO)_3CH_2SiHMe_2 \xrightarrow{UV} \underset{Me\ Me}{OC\text{-}W(Cp)(H)(CO)\text{-}Si\text{-}CH_2} \quad (7)$$

$$Cp^*Fe(CO)_2CH_2SiHMe_2 \xrightarrow{UV} \begin{array}{c} Cp^* \\ | \\ OC^{\prime\prime\prime}Fe\diagdown_{}H \\ Me\diagup Si\diagdown C\diagdown H \\ Me \end{array} \qquad (8)$$

In 1988, Tilley and coworkers first succeeded in the synthesis of a ruthenium–silene complex by the reaction of $Cp^*Ru(PR_3)Cl$ with a Grignard reagent, and the structure was confirmed by X-ray crystallography (Eq. 9) [10]. They later synthesized an iridium complex in a similar manner (Eq. 10) [11]. Berry synthesized a tungsten–silene complex by treatment of a tungsten-chloride complex with Mg (Eq. 11) [12]. Although reactions of transition metal–silene complexes with MeI, HX, and MeOH have been reported, little is known about their reactivities [11, 12].

$$Cp^*Ru(PR_3)Cl \xrightarrow{ClMgCH_2SiR'_2H} Cp^*Ru(PR_3)CH_2SiR'_2MeH \longrightarrow \begin{array}{c} Cp^* \\ | \\ H^{\prime\prime\prime}Ru-SiR'_2 \\ Ln \end{array} \qquad (9)$$

1a R = Cy R' = Me
1b R = ^{i}Pr R' = Ph
1c R = Cy R' = Ph

$$Cp^*(PMe_3)IrMeCl \xrightarrow{ClMgCH_2SiPh_2H} \underset{\mathbf{1d}}{\begin{array}{c} Cp^* \\ | \\ Ir^{\prime\prime\prime}SiPh_2 \\ Me_3P \end{array}} \begin{array}{c} \xrightarrow{MeI} \begin{array}{c} Cp^* \\ | \\ Ir^{\prime\prime\prime}I \\ Me_3P \quad CH_2SiMePh_2 \end{array} \\ \xrightarrow{MeOH} \begin{array}{c} Cp^* \\ | \\ Ir^{\prime\prime\prime}H \\ Me_3P \quad CH_2Si(OMe)Ph_2 \end{array} \end{array} \qquad (10)$$

$$\begin{array}{c} CpW^{\prime\prime\prime}Cl \\ CH_2SiMe_2Cl \end{array} \xrightarrow{Mg} \underset{\mathbf{1f}}{Cp_2W\diagdown_{SiMe_2}} \begin{array}{c} \xrightarrow{MeOH} Cp^*W\diagup^{H}_{CH_2SiMe_2OMe} \\ \xrightarrow{HX} Cp^*W\diagup^{SiMe_3}_{X} \end{array} \qquad (11)$$

Very recently, Berry and coworkers reported the formation of a ruthenium–silene complex from $(PMe_3)_4RuHSiMe_3$ and BPh_4, and they observed the interconversion between silene and a 16-electron ruthenium silyl complex (Eq. 12) [13a].

$$\begin{array}{c} P \\ P^{\prime\prime}\diagdown | \diagup SiMe_3 \\ Ru \\ P\diagup | \diagdown H \\ P \end{array} BPh_4 \longrightarrow \begin{array}{c} P \\ P^{\prime\prime}\diagdown | \diagup SiMe_3 \\ Ru \\ \diagup | \diagdown H \\ P \end{array} \rightleftharpoons \begin{array}{c} Me \\ Si\diagup Me \\ H^{\prime\prime}C\diagdown \diagup H \\ Ru \\ H\diagup P \diagdown P \\ P \end{array} \longleftrightarrow \begin{array}{c} Me\diagdown \diagup Me \\ Si \\ \diagup \diagdown H \\ Ru \\ H^{\prime\prime}\diagup P \diagdown P \\ H \quad P \end{array} \qquad (12)$$

P=PMe$_3$

Tilley reported silene–silylene rearrangement in a cationic iridium complex (Eq. 13) [13b].

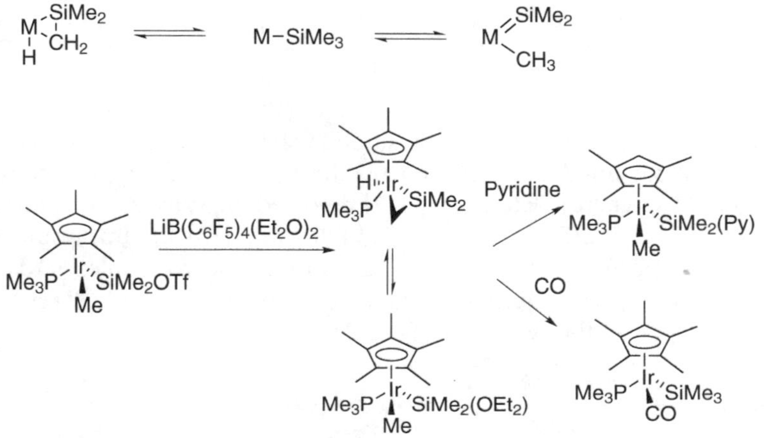

$$\underset{H}{\overset{SiMe_2}{M-CH_2}} \rightleftharpoons M-SiMe_3 \rightleftharpoons \underset{CH_3}{\overset{SiMe_2}{M}}$$

(13)

To examine the reactivity of a zirconium–silicon bond, Mori et al. synthesized a zirconium–silicon complex from Cp_2ZrCl_2 and $Ph_2{}^tBuSiLi$ (1 equiv.) and investigated the reactivity of this complex [14]. Furthermore, they found that the reaction of Cp_2ZrCl_2 with 2 equivalents of $Me_2PhSiLi$ gave disilylzirconocene, which was easily converted into a zirconium–silene complex via β-hydrogen elimination (Scheme 1) [15]. This is the first example of the formation of an early transition metal–silene complex. In this chapter, the reactivity of the zirconium–silene complex is described.

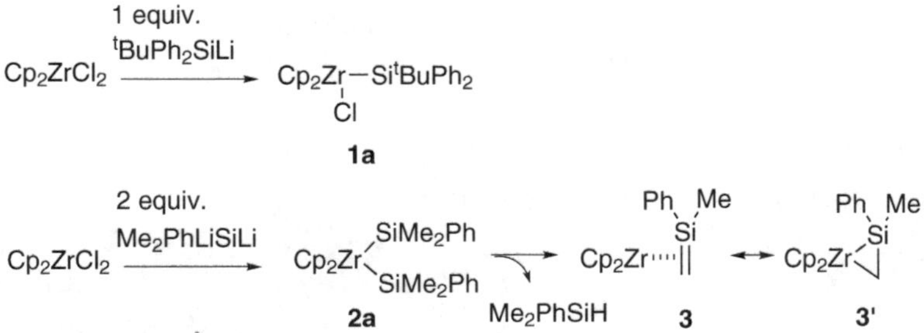

Scheme 1 Reaction of Cp_2ZrCl_2 with R_3SiLi

1.1
Reaction of Alkyne with Cp_2ZrCl_2 and $Ph_2{}^tBuSiLi$

The reactivity of a silyl–zirconium complex is interesting because an unsaturated bond would be inserted into the silyl–zirconium bond to provide an alternative zirconium complex. It has zirconium–carbon and silyl–carbon bonds and it is expected that novel carbon–carbon bond-forming reactions would be developed.

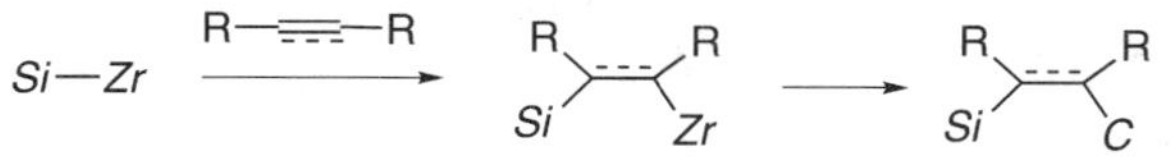

$$\text{Si-Zr} \xrightarrow{R\equiv R} \underset{Si\quad Zr}{R\diagup\!\!\!\diagdown R} \longrightarrow \underset{Si\quad C}{R\diagup\!\!\!\diagdown R} \quad (14)$$

These are only few reports on the synthesis and reactivity of complexes having a zirconium–silicon bond. Lappert reported the synthesis of complex 1b from Cp_2ZrCl_2 and Ph_3SiLi [16]. Later, Tilley [17] reported the synthesis of complex 1c from Cp_2ZrCl_2 and $Al(SiMe_3)_3 \cdot OEt_2$ [17a]. Takahashi [18] and Buchwald [19] independently reported the formation of zirconium–silicon complexes 1d and 1e by treatment of zirconocene coordinated by olefin with silane. Berry [20] and Xue [21] reported the syntheses of complexes 1f and 1g (Scheme 2).

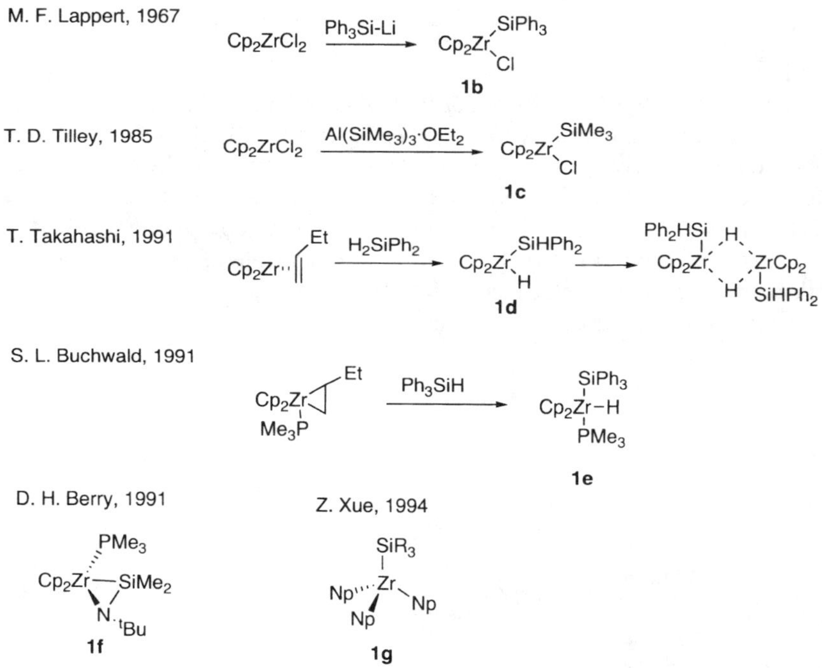

Scheme 2 Preparation of zirconium–silicon complexes

Little is known about the reactivities of complexes having zirconium–silicon bonds. Reaction of 1b with hydrogen chloride afforded triphenylsilane (Scheme 3) [16a]. The insertion of carbon monoxide or isocyanide into a zirconium–silicon bond of 1c gave silaacylzirconium complex 4 or iminosilylzirconium complex 6 [17b,c]. As for carbon–carbon multiple bonds, ethylene can be inserted into a zirconium–silicon bond of 1h' [17g], but other multiple bonds such as alkene and alkyne could not be inserted into a zirconium–silicon bond [17b,c].

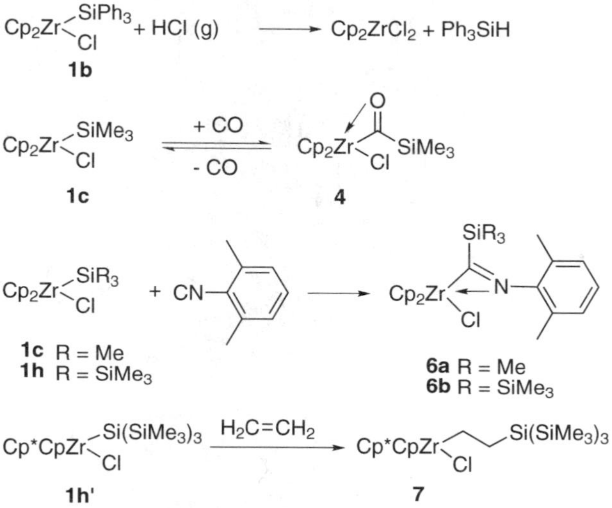

Scheme 3 Reactivity of Cp$_2$Zr(SiR$_3$)Cl

Zirconium–silyl complex **1a** was prepared from Cp$_2$ZrCl$_2$ and LiSi^{t}BuPh$_2$ [22], but the isolation of zirconium–silyl complex **1a** was unsuccessful. Thus, the insertion of isonitrile into a zirconium–silyl bond was carried out in situ to give air- and moisture-stable η^2-iminosilaacyl complex **6c** in 71% yield after column chromatography on silica gel (m.p. 193–194 °C, from hexane-CH$_2$Cl$_2$, Scheme 4). The structure of **6c** was confirmed by X-ray diffraction. η^2-Iminosilaacyl complex **6c** was treated with dry HCl in benzene to give formidoylsilane **9** in 53% yield.

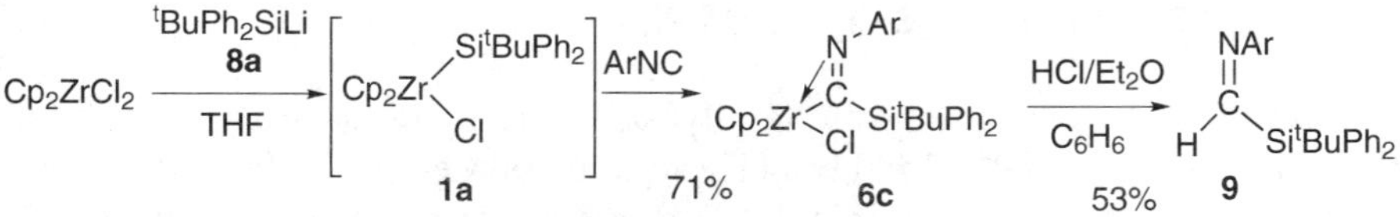

Scheme 4 Reaction of Cp$_2$ZrCl$_2$ with ^{t}BuPh$_2$SiLi

Treatment of η^2-iminosilaacyl complex **6c** with LiEt$_3$BH gave azazirconacyclopropane **10**, which was hydrolyzed to give (silylmethyl)aniline in 82% yield (Scheme 5). Treatment of **10** with 4-octyne gave alkene **12** in 73% yield after hydrolysis. Presumably, the insertion of alkyne into the carbon–zirconium bond of **10** gives silazirconacyclopentene **11**. When CuCl and allyl chloride were added to a THF solution of silazirconacyclopentene **11**, tetrasubstituted alkene

13 was obtained via transmetalation. These results indicated that two functional groups, a formyl group and an allyl group, could be introduced onto alkyne in a highly stereoselective manner in a one-pot reaction.

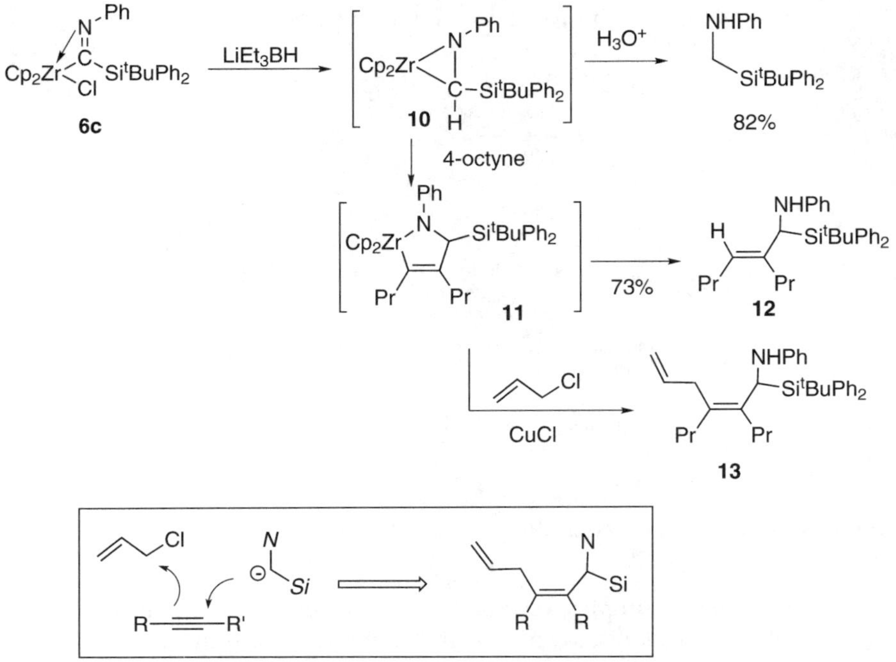

Scheme 5 Treatment of silaacylzirconium complex with LiEt$_3$BH

2
Formation of Zirconium–Silene Complex

2.1
Reaction of Alkyne with Cp$_2$ZrCl$_2$ and Me$_2$PhSiLi

To examine the reactivity of the silyl–zirconium bond, Mori et al. used Me$_2$PhSiLi instead of Ph$_2$^{t}BuSiLi. A THF solution of Me$_2$PhSiLi **8b** (1 equiv.) was added to a THF solution of Cp$_2$ZrCl$_2$ (1 equiv.) and diphenylacetylene **14a** (1 equiv.) at –78 °C, and the solution was stirred at room temperature for 3 h. After hydrolysis of the reaction mixture, vinylsilane **15a** was obtained in 36% yield along with the starting material **14a** in 40% yield. When the reaction mixture was treated with D$_2$O, compound **15a-D$_2$** having two deuteriums was obtained. One deuterium was introduced at the vinylic position and the other was incorporated into the methyl proton on the silicon (39% yield, each D content; quant.). If insertion of alkyne **14a** into the zirconium–carbon bond of **1b** occurs, vinylsilane **15a** should be formed, but in this case only one deuterium should be introduced at the vinylic position in the molecule. It cannot be explained at this stage why two deuteriums were introduced into the molecule (Scheme 6).

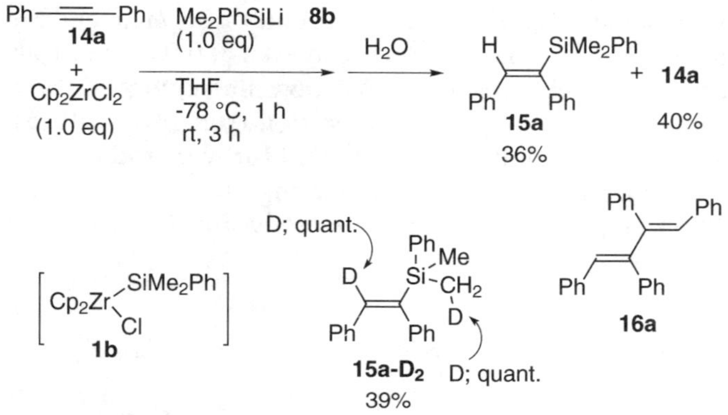

Scheme 6 Reaction of Cp$_2$ZrCl$_2$ and Me$_2$PhSiLi in the presence of **14a**

The reaction was carried out under various conditions to improve the yield of the vinylsilane **15a** (Table 1). When 2 equivalents of Me$_2$PhSiLi **8b** to Cp$_2$ZrCl$_2$ were used for this reaction, the yield of **15a** increased to 68% along with dimeric compound **16a** in 6% yield (run 2), but a 1:1 molar ratio of Cp$_2$ZrCl$_2$ and Me$_2$PhSiLi did not affect the yield of **15a** (run 3). The yield was improved to

Table 1 Reaction of Cp$_2$ZrCl$_2$, Me$_2$PhSiLi, and **14a**[a]

Run	Cp$_2$ZrCl$_2$ (equiv.)	Me$_2$PhSiLi (equiv.)	Temp. (°C)	Yield of **15a** (%)
1	1	1	rt	36[b]
2	1	2	rt	68
3	2	2	rt	59
4	1.5	3	rt	76
5	1.5	3	rt	78[c]
6	1.5	3	40	74
7	2	4	rt	82
8	1.5	3	0	66
9	1.5	3	0	74[d]
10	1.5	3	rt	79[e]

[a] To a THF solution of Cp$_2$ZrCl$_2$ and **14a** was added **8b** in THF at –78 °C, the solution was stirred at –78 °C for 1 h, and then the solution was stirred at ambient temperature for 3 h.
[b] **14a** was recovered in 40% yield.
[c] D$_2$O was added to there action mixture and **15a-D$_2$** was obtained.
[d] Reaction time: 6 h.
[e] Toluene was used as the solvent and a THF solution of **8b** was added.

76% when 1.5 equiv. of Cp$_2$ZrCl$_2$ and 3 equiv. of Me$_2$PhSiLi were used (run 4). When the reaction was carried out under the same conditions and deuterolysis was carried out again, the same product **15a-D$_2$** was obtained (run 5). A higher reaction temperature did not affect the yield of **15a** (run 6). The yield of **15a** was increased to 82% when excess amounts of Cp$_2$ZrCl$_2$ and Me$_2$PhSiLi **8b** were used (run 7). The reaction proceeded even at 0 °C (run 8), although the yield was slightly lower, but a longer reaction time improved the yield of **15a** (run 9). Toluene can be used for this reaction (run 10). In all cases, a small amount (less than 8%) of dimeric compound **16a** was produced.

Various alkynes **14** gave the corresponding vinylsilanes **15** in high yields (Scheme 7). The electron-donating group on the aromatic ring caused a slight

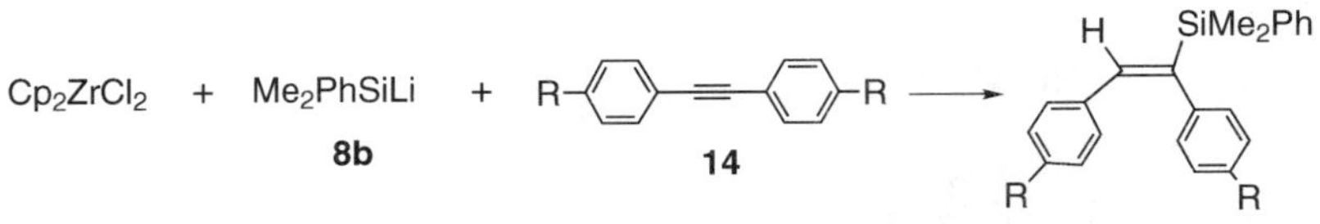

Scheme 7 Synthesis of various vinylsilanes **15**

15b R=OMe 81%
15c R=Me 84%
15d R=CF$_3$ 66%

increase in the yield of the desired vinylsilane **15**. On the other hand, when 3-hexyne **17a** was reacted with Cp$_2$ZrCl$_2$ (1 equiv.) and Me$_2$PhSiLi (2 equiv.), surprisingly, unexpected allylsilane **18a** was obtained, although the yield was low (Table 2, run 1; Scheme 8). When the reaction mixture was treated with D$_2$O, two deuteriums were also incorporated, one at the vinylic position and the other on the silicon of **18a-D$_2$**. (D contents: 68 and 85%, respectively). A higher reaction temperature increased the yield of the desired compound **18a** (Table 2, run 2), but the addition of PPh$_3$ as a ligand did not give a good result (run 3). Allylsilane **18b** was also obtained when 4-octyne was used for this reaction

Table 2 Synthesis of allylsilane[a]

Run	R	Alkyne	Temp. (°C)	Product	Yield (%)
1	Et	17a	rt	18a	15
2	Et	17a	40	18a	41
3	Et	17b	40	18b	29[b]
4	Pr	17b	40	18b	25
5	Pr	17b	70	18b	35

[a] To a THF solution of Cp$_2$ZrCl$_2$ and **17** was added **8b** in THF at −78 °C, the solution was stirred at −78 °C for 1 h, and then the solution was stirred at ambient temperature for 3 h.
[b] PPh$_3$ (1.0 equiv.) was added.

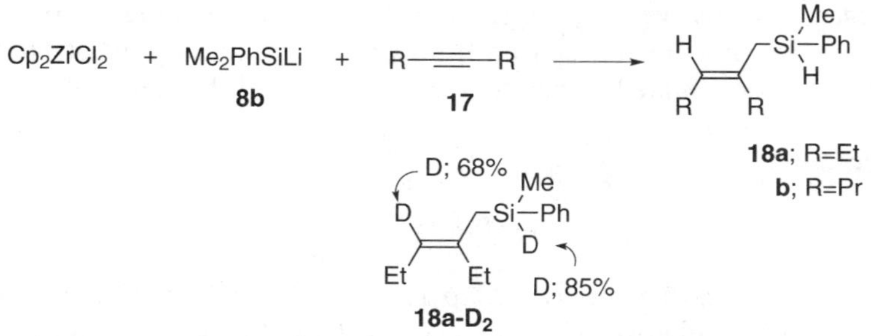

Scheme 8 Reaction with alkyne having an alkyl group

(run 4), and in this case, the yield of **18b** increased when the reaction was carried out at 70 °C (run 5).

As a source of alkyne, enyne **19** was used in this reaction to afford two inseparable regioisomers of vinylsilanes **20a** and **20b** in 54% yield in a ratio of 3:1 (Scheme 9).

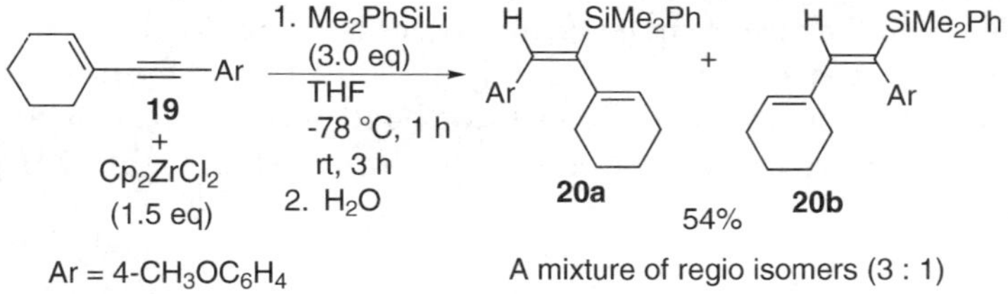

Scheme 9 Reaction of zirconium–silene complex and enyne

2.2
Possible Reaction Course for the Formation of Zirconium–Silene Complex

A possible reaction course was considered (Scheme 10). The reaction of Cp_2ZrCl_2 with $Me_2PhSiLi$ should give chlorosilyl zirconocene **1b**. Insertion of alkyne **14a** into the zirconium–silicon bond of **1b** should give **21**. Hydrolysis of **21** gives vinylsilane **15a**. However, in this case, one deuterium should be incorporated into **15a** to give **15a-D**, not **15a-D$_2$**.

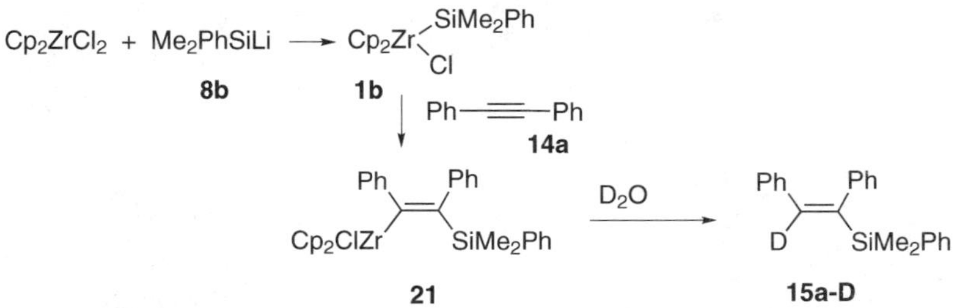

Scheme 10 Possible reaction course

Thus, an alternative reaction mechanism must be considered. It is known that dibutylzirconocene, prepared from Cp_2ZrCl_2 and 2 equiv. of BuLi, gives zirconocene coordinated by a butene ligand (Negishi's reagent, Eq. 15) [23].

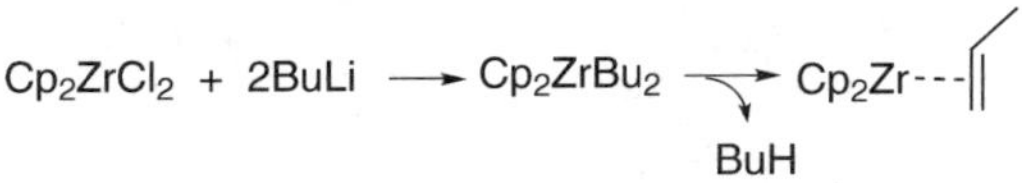

$$Cp_2ZrCl_2 + 2BuLi \longrightarrow Cp_2ZrBu_2 \xrightarrow{\text{BuH}} Cp_2Zr\text{---}|| \qquad (15)$$

Therefore, complex **1b** formed from Cp_2ZrCl_2 and $Me_2PhSiLi$ was further reacted with $Me_2PhSiLi$ to give disilylzirconocene **2a**. Then, **2a** would be converted into zirconium–silene complex **3** or silazirconacyclopropane **3'** by β-hydrogen elimination from **2a**. The insertion of alkyne **14a** into the zirconium–silicon bond of silazirconacyclopropane **3'** gives silazirconacyclopentene **22a**. Thus, in this case, deuterolysis of silazirconacyclopentene **22a** gives **15a-D$_2$**, which has two deuteriums, at the vinylic position and at the methyl carbon of the silyl group. On the other hand, the insertion of dialkyl alkyne **17a** into the carbon–zirconium bond of **3'** gives silazirconacyclopentene **23a**, whose deuterolysis gives **18a-D$_2$**. Deuteriums were incorporated at the vinylic position and on the silicon (Scheme 11). In the case of enyne **19**, the double bond on the alkyne would have the same role as that of the aromatic ring on alkyne **14**, and the insertion of the alkyne part of **19** into the zirconium–silicon bond of zirconium–silene complex **3'** occurs to give silazirconacyclopentene and hydrolysis of it gives **20a** and **20b**.

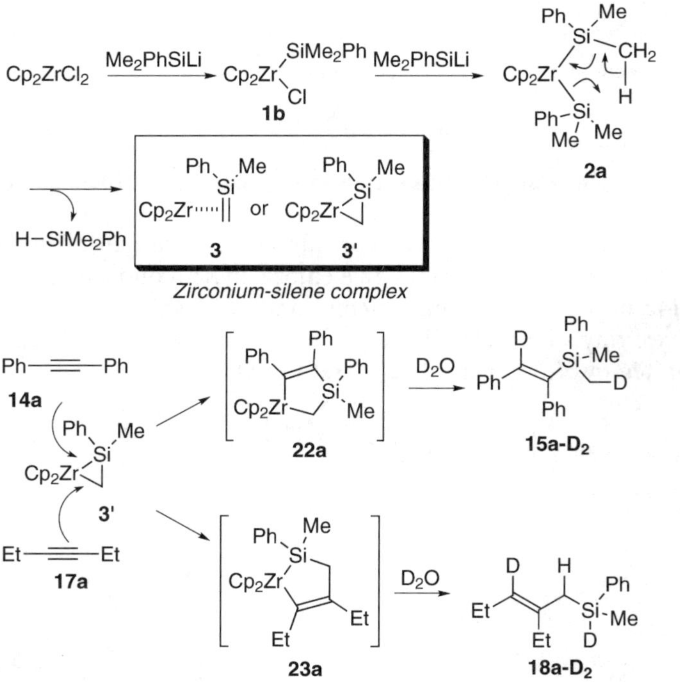

Scheme 11 Alternative possible reaction course

The reasons why alkynes 14 and 19 having aryl or vinyl groups, respectively, insert into the zirconium–silicon bond of 3', whereas the alkyne 17 having alkyl groups inserts into the carbon–zirconium bond of 3', are still not clear. Presumably, electronic factors are important for the insertion reaction.

Although silene is known to be unstable, this zirconium–silene complex is easily generated from disilylzirconocene 2a by β-hydrogen elimination in situ. However, the isolation of the zirconium–silene complex is difficult, and it immediately reacts with alkyne to form the azazirconacyclopentene.

2.3
Confirmation of the Reaction Mechanism

To confirm the generation of a zirconium–silene complex, the reaction of Cp$_2$ZrCl$_2$ and Me$_2$PhSiLi **8b** in the presence of bis-4-methoxyphenylacetylene **14b** was monitored by ^{1}H NMR (Scheme 12).

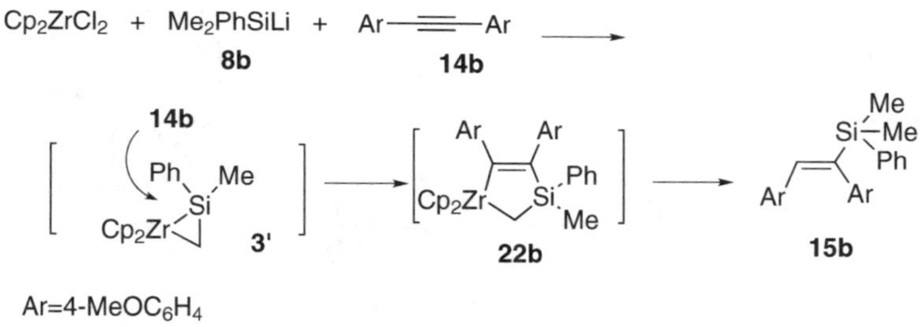

Ar=4-MeOC$_6$H$_4$

Scheme 12 Reaction of alkyne **14b** with silene **3'**

At first, the NMR spectrum of a THF-d_8 solution of Cp$_2$ZrCl$_2$ and **14b** was measured. The Cp protons of Cp$_2$Zr$_2$Cl$_2$ appeared at δ 6.48. Then a THF solution of Me$_2$PhSiLi was added to this solution at –78 °C, and the ^{1}H NMR spectrum was measured at room temperature. Cp protons appeared at δ 5.07, and a Si-H proton of Me$_2$PhSiH was clearly shown at δ 4.42, whose δ value was confirmed by authentic Me$_2$PhSiH in THF-d_8 (0 min). Then the solution was allowed to stand at room temperature and was monitored by ^{1}H NMR. After 7 min, new peaks appeared at δ 6.30 and 6.17 (Fig. 1, Chart 1). As a function of time, the peak at δ 5.07 decreased whereas the peaks at δ 6.30 and 6.17 increased (Chart 2). After 4.3 h, the peak at δ 5.07 had disappeared (Chart 3). When HCl-Et$_2$O was added to this solution, a single peak of Cp$_2$ZrCl$_2$ appeared at δ 6.50. From the reaction mixture in the NMR tube, **15b** was isolated in 69% yield. This means that the Cp protons of δ 6.30 and 6.17 in Chart 3 are those of silazirconacyclopentene **22b**.

Subsequently, the reaction of Cp$_2$ZrCl$_2$ and Me$_2$PhSiLi in the absence of the alkyne **14b** was monitored by ^{1}H NMR. The results are shown in Charts 4–6 (Fig. 2). After addition of Me$_2$PhSiLi to the solution of Cp$_2$ZrCl$_2$ in THF-d_8, the ^{1}H NMR spectrum was immediately measured. Cp protons appeared at δ 6.11,

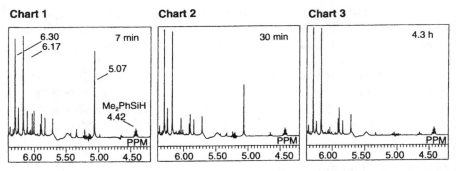

Fig. 1 1H NMR spectra of the reaction of Cp$_2$ZrCl$_2$ and **8b** in the presence of **14b**

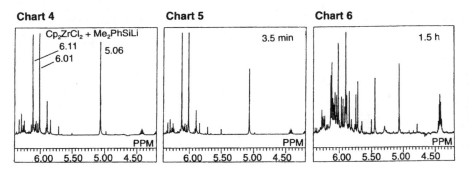

Fig. 2 ^{1}H NMR spectra of the reaction of Cp$_2$ZrCl$_2$ and **8b**

6.01, and 5.06 (Chart 4). After 3.5 min, there was no change in the spectrum (Chart 5). However, after 1.5 h, many peaks had appeared on the NMR spectrum (Chart 6).

On the other hand, a THF solution of **8b** was added to Cp$_2$ZrCl$_2$ in THF-d_8 and the ^{1}H NMR spectrum was measured (Chart 7, same conditions as those for Chart 4, Fig. 3). Then alkyne **14b** was immediately added to the NMR tube.

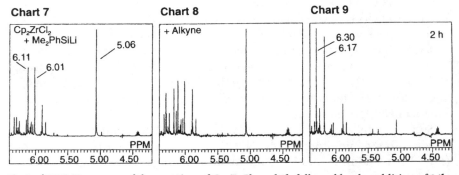

Fig. 3 ^{1}H NMR spectra of the reaction of Cp$_2$ZrCl$_2$ and **8b** followed by the addition of **14b**

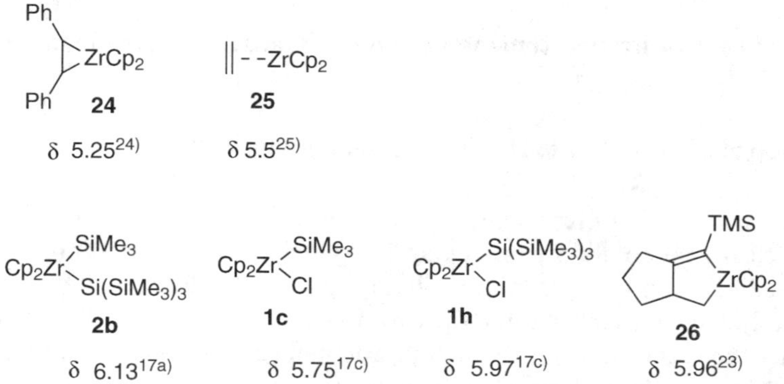

Synthesis and Reactivity of Zirconium–Silene Complex 55

[Structures:]

Ph₂C=ZrCp₂ 24   δ 5.25[24]

||--ZrCp₂   25   δ 5.5[25]

Cp₂Zr(SiMe₃)(Si(SiMe₃)₃) 2b   δ 6.13[17a]

Cp₂Zr(SiMe₃)(Cl) 1c   δ 5.75[17c]

Cp₂Zr(Si(SiMe₃)₃)(Cl) 1h   δ 5.97[17c]

Bicyclic TMS-ZrCp₂ 26   δ 5.96[23]

Fig. 4 The δ values of Cp protons

New peaks appeared at δ 6.30 and 6.17 (Chart 8). After 2 h, the peaks of δ 5.06, 6.01, and 6.11 had disappeared (Chart 9). Chart 9 was almost the same as Chart 3. From this reaction mixture, **15b** was isolated in 33% yield.

The δ values of typical Cp protons of zirconacycles and zirconocenes in the literature are shown in Fig. 4 [17a,c, 23–25]. Although the chemical shift of Cp protons of a zirconium–silene complex [Zr(II)] or silazirconacyclopropane is not known, it has been reported that the chemical shifts of Cp protons of zirconacyclopropane **24** (δ 5.25) [24] or zirconocene **25** [Zr(II), δ 5.50] [25] coordinated by an ethylene ligand are greater than those of Cps of Zr(IV) complexes. Thus, the peak of δ 5.06 appearing at a higher chemical shift should be that of the Cp peak of Zr(II) and it would be the Cp proton of zirconium(II)-silene complex **3** or **3'**. This is supported by the fact that the peak of δ 5.06 decreased when alkyne **14b** was added to the THF-d_8 solution, and new peaks appeared at δ 6.30 and 6.17 (Charts 8 and 9). This indicates that zirconium–silene complex **3** or **3'** was converted into silazirconacyclopentene **22b** in the presence of the alkyne **14b**, and vinylsilane **15b** was isolated from the reaction mixture in the NMR tube after the usual workup. Therefore, the Cp peaks at δ 6.30 and 6.17 in Charts 3 and 9 are those of silazirconacyclopentene **22b**, because the chemical shift of Cp peaks of zirconacyclopentene **26** is known to be δ 5.96 [24]. Since silene is very unstable, it decomposed after 1.5 h in the absence of alkyne (Chart 6). The Cp peaks of δ 6.11 and 6.01 (Chart 4) are thought to be those of disilylzirconocene **2a** or chlorosilylzirconocene **1b**, by comparison with that of disilylzirconocene **2b** (δ 6.13) [17a] and those of chlorosilylzirconocenes **1c** or **1h** (δ 5.75 [17c] or 5.97 [17c]). These results strongly suggested that silene is formed from disilylzirconocene **2a**, generated from Cp₂ZrCl₂ and 2 equiv. of Me₂PhSiLi **8b**, and it reacted with **14b** to give silazirconacyclopentene **22b**, which afforded vinylsilane **15b** after hydrolysis (Scheme 11).

2.4
Reactivity of Silazirconacyclopentene Formed from a Zirconium–Silene Complex

2.4.1
Insertion of Isocyanide into Silazirconacyclopentene

The reaction of zirconium–silene complex 3 with alkyne was carried out, and it was clear that an alkyne having aryl or vinyl groups inserts into the zirconium–silicon bond of zirconium–silene complex 3′, while an alkyne possessing two alkyl groups inserts into the carbon–zirconium bond of 3′ (Scheme 11).

Thus, the reaction of silazirconacyclopentene 22 formed from silene–zirconium complex 3′ and alkyne 14 with isonitrile was carried out [15b]. To a THF solution of Cp$_2$ZrCl$_2$ (1.5 equiv.) and diphenylacetylene 14a (1 equiv.) was added Me$_2$PhSiLi (3 equiv.) at –78 °C, and the solution was stirred at the same temperature for 1 h and then at room temperature for 3 h (Scheme 13). To this solution was added tert-butyl isocyanide (2.1 equiv.), and the solution was stirred at room temperature overnight. Surprisingly, treatment of the reaction mixture with HCl-Et$_2$O gave imino–zirconium complex 27a, but neither 29a nor 30a, in 57% yield based on alkyne 14a. When the reaction mixture was quenched with 20% DCl-D$_2$O, 27a-D was obtained in 68% yield (D content: quant.). The intermediate would be iminosilazirconacyclohexene 28a, and the carbon–zirconium bond remained unchanged due to the strong coordination of nitrogen in 27a to zirconium.

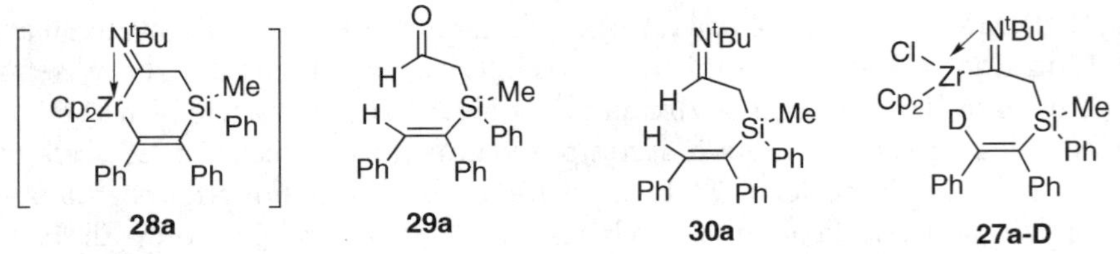

Scheme 13 Reaction of silazirconacyclopentene and ^{t}BUNC

Various imino–zirconium complexes 27 were obtained from Cp$_2$ZrCl$_2$ (1.5 equiv.), Me$_2$PhSiLi (3 equiv.), and the corresponding diarylalkynes 14a–d (1 equiv.) by a one-pot reaction in good to moderate yields (Table 3). In this reaction, the electron-donating group on the aromatic ring gave good results. Since compound 27d was crystallized, its structure was confirmed by X-ray crystallography [15b]. To confirm the formation of iminosilazirconacyclohex-

Table 3 Synthesis of imino–zirconium complex

Run	Ar	Alkyne	Product	Yield (%)
1	Ph	14a	27a	57
2	4-MeOC$_6$H$_4$	14b	27b	62
3	4-MeC$_6$H$_4$	14c	27c	68
4	4-CF3C$_6$H$_4$	14d	27d	27

NMR-Experiment for Synthesis of Iminosilazirconacyclohexene

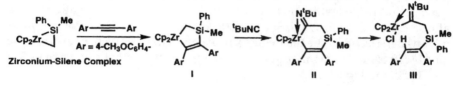

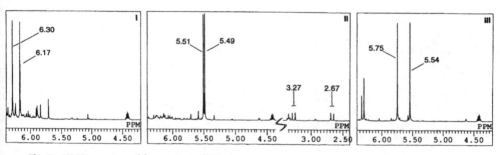

Fig. 5 NMR spectra of formation of **27b**

ene **28**, the reaction was monitored by ^{1}H NMR spectra. The spectra of **27b** are shown in Fig. 5.

To a solution of Cp$_2$ZrCl$_2$ (15.9 mg, 54.4 μmol) and di-4-methoxyphenylacetylene **14b** (8.5 mg, 35.7 μmol) in THF-d_8 (0.8 mL) was added Me$_2$PhSiLi (0.72 M in THF, 0.15 mL, 108 μmol) in an NMR tube under argon atmosphere, and the solution was allowed to stand at −78 °C. Then the solution was monitored at room temperature by ^{1}H NMR spectra. As a function of time, the Cp proton signals of silazirconacyclopentene **22b** at δ 6.17 and 6.30 increased (Chart I). After 3.5 h, tert-butyl isocyanide (10.0 μL, 88.4 μmol) was added. New Cp proton peaks appeared at δ 5.49 and 5.51, and methylene peaks appeared at δ 2.67 (d, J=12.4 Hz, 1H) and δ 3.27 (d, J=12.4 Hz, 1H), respectively (Chart II). When HCl-Et$_2$O was added to this solution, the Cp protons were changed to δ 5.75 and 5.54, whose peaks agreed with those of imino–zirconium complex **27b** (Chart III). The Cp protons at δ 5.49 and 5.51 (10H) and methylene protons at δ 2.67 (d, J=12.4 Hz, 1H) and δ 3.27 (d, J=12.4 Hz, 1H) as an AB quartet in Chart II indicate that iminosilazirconacyclohexene **28b** is formed.

2.4.2
Insertion of Carbon Monoxide

Next, an attempt at insertion of carbon monoxide into silazirconacyclopentene 22b prepared from silene and diarylalkyne was made. A THF solution of silazirconacyclopentene 22b, which was prepared from alkyne 14b, Cp$_2$ZrCl$_2$, and Me$_2$PhSiLi in THF, was stirred under carbon monoxide (1 atm) at room temperature overnight (Scheme 14). From the ^{1}H NMR, mass, and IR spectra of the product, it was clear that the desired carbonylation product 32 was not formed. To have more insight into the structure of the product, the reaction mixture was quenched with D$_2$O: two deuteriums were incorporated at the vinyl and at the acetyl methyl carbons of the product. Hydrogenation of 31b afforded bibenzyl 33b. On the basis of these results, the structure of the product was determined to be the methyl silyl ketone 31b. Treatment of alkynes 14a and 14c in a similar manner gave the corresponding methyl silyl ketones 31a and 31c in moderate yields.

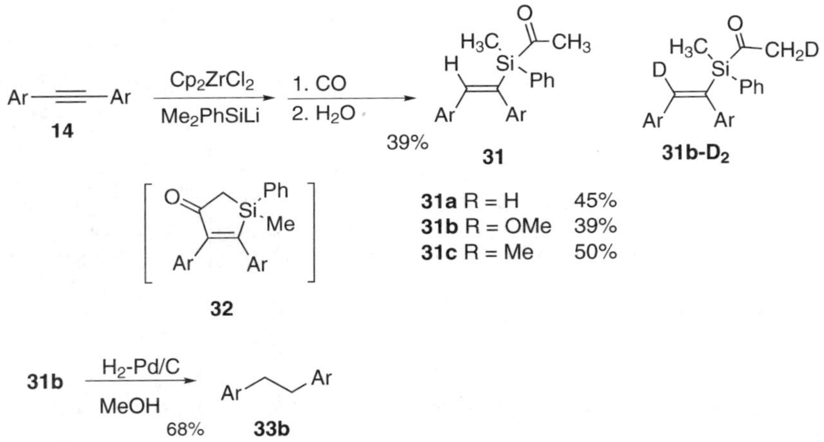

Scheme 14 Reaction of silazirconacyclopentene with carbon monoxide

The possible reaction mechanism for the formation of 31 is shown in Scheme 15. Insertion of alkyne 14 into silazirconacyclopropane 3' gives silazirconacyclopentene 22. Then, insertion of carbon monoxide into the carbon–zirconium bond in silazirconacyclopentene 22 gives silazirconacyclohexenone 34, whose carbonyl oxygen would coordinate to zirconium metal. Then the zirconium carbon bond migrates onto silicon to afford oxazirconacyclohexene 36 via 35 [26]. Deuterolysis of 36 would afford 31-D$_2$, which has two deuteriums.

It is interesting to note that hydrolysis of iminosilacylcyclohexene 28 gave iminozirconium complex 27, while silazirconacyclohexenone 34 was converted into oxazirconacycle 36, which when treated with H$_2$O gave the methyl silyl ketone 31 (Scheme 16). The differences between the reactivities of iminosilazirconacyclohexene 28 and silazirconacyclohexenone 34 would be due

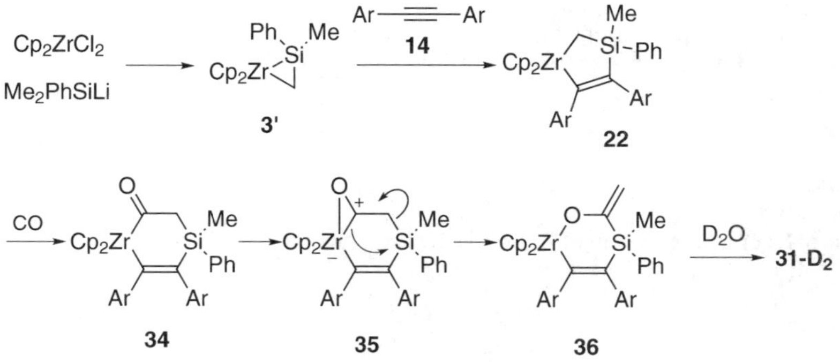

Scheme 15 Possible reaction mechanism

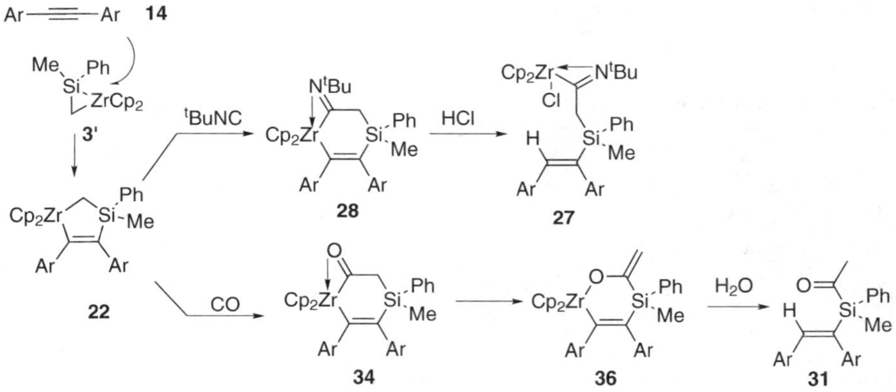

Scheme 16 Difference in reaction course

to the strong coordination of the carbonyl oxygen in **34** to zirconium metal compared with that of the imino nitrogen in **28** to zirconium metal. These substituents were introduced on the carbon of silene via silazirconacyclopentene **22**.

2.4.3
Transmetalation of Zirconium to Copper of Silazirconacyclopentene

It is known that the transmetalation reaction from zirconium to copper is a useful tool for the formation of a new carbon–carbon bond. Schwartz reported the first transmetalation from zirconium to copper [27], while Lipschutz [28] and Takahashi [29] used it for synthetic purposes. Thus, the transmetalation reaction of silazirconacyclopentene to copper was investigated. To a THF solution of CuCl (2 equiv. to Cp$_2$ZrCl$_2$) and allyl chloride was added a THF solution of silazirconacyclopentene **22a**, generated from Cp$_2$ZrCl$_2$, alkyne **14a**, and Me$_2$PhSiLi **8b**, and the solution was stirred at room temperature for 18 h. After the usual workup, the bis-allylated compound **37a** was obtained in 76% yield

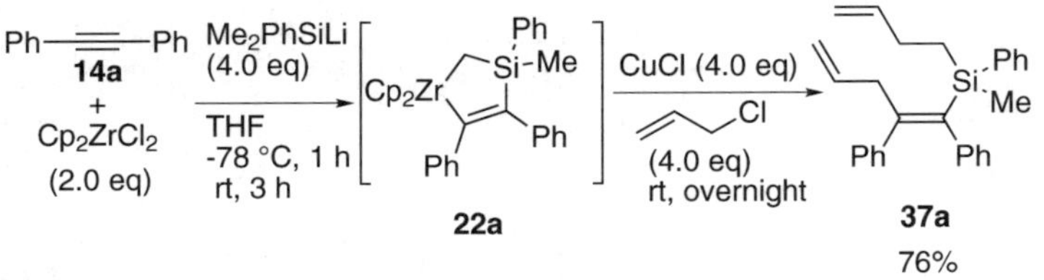

Scheme 17 Transmetalation of Zr on silazirconacyclopentene with CuCl

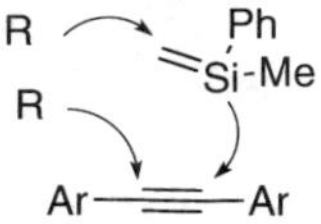

Fig. 6 Introduction of allyl and butenylmethylphenyl silyl groups

Table 4 Transmetalation of 22 to copper

run	Ar		Product	Yield (%)
1	Ph	 ⩘Cl	37a	76
2	4-CH$_3$OC$_6$H$_4$	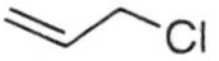 ⩘Cl	37b	72
3	4-CH$_3$C$_6$H$_4$	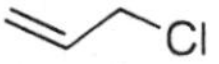 ⩘Cl	37c	76
4	Ph	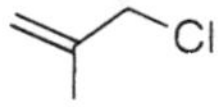 ⩘Cl	37d	66

(Scheme 17). The results indicated that the two allyl groups were simultaneously introduced on the alkynyl carbon and on the methyl group on the silicon moiety. As an overall result, two different substituents were introduced, in a one-pot reaction, on two alkynyl carbons. One is the allyl group whereas the other is the butenylmethylphenyl silyl group (Fig. 6).

Transmetalations of various zirconacycles **22a–d** to copper were carried out in the presence of allyl halide, and the results are shown in Table 4. In each case, bis-allylated compounds **37a–d** were obtained in high yields.

3
Perspective

A zirconium–silene complex was obtained during the course of our study on the synthesis of a zirconium–silyl complex. The zirconium–silene complex is formed from disilylzirconocene **2a**, generated from Cp$_2$ZrCl$_2$ and 2 equivalents of Me$_2$PhSiLi **8b**. Zirconium–silene complex **3** should be in a state of equilibrium with zirconacyclopropane **3'**. The insertion of diaryl alkyne **14** or vinyl alkyne

19 into the zirconium-silicon bond of silazirconacyclopropane 3' gives silazirconacyclopentene 22. On the other hand, insertion of dialkyl alkyne 17 into the zirconium-carbon bond of zirconacyclopropane 3' gives silazirconacyclopentenes 23. As a result, the silyl group or the silylmethyl group could be introduced on the alkynyl carbon. It is interesting to note that the methyl group on the silicon of the silazirconacyclopentene could react with various electrophiles such as protons, deuteriums, isocyanide, carbon monoxide, and allyl halides.

Since zirconium-silene complex 3 could be easily synthesized in situ, the insertion of various compounds into the zirconium-silicon or zirconium-carbon bonds of silazirconacyclopropane 3' will be investigated in a future study, and it is expected that a novel carbon-carbon or carbon-silicon bond-forming reaction will be developed.

References

1. For reviews, see: (a) Patai S, Rapopport Z (eds) (1989) The chemistry of organic silicon compounds, part 2. Wiley, New York, chap 17; (b) Zybill CE, Liu C (1995) Synlett 687; (c) Raabe G, Michl J (1985) Chem Rev 85:419; (d) Gusel'nikov LE, Nametkin NS (1979) Chem Rev 79:529
2. Gusel'nikov LE, Flowers MCJ (1967) Chem Soc Chem Commun 864
3. (a) Gusel'nikov LE, Nametkin NS, Vdovin VM (1975) Acc Chem Res 8:12; (b) Nametkin NS, Gusel'nikov LE, Ushakova RL, Vdovin VM (1971) Dokl Akad Nauk SSSR 201:1365
4. (a) Brook AG, Abdesaken F, Gutekunst B, Gutekunst G, Kallury RK (1981) J Chem Soc Chem Commun 191; (b) Brook AG, Nyburg SC, Abdesaken F, Gutekunst B, Gutekunst G, Kallury RKMR, Poon YC, Chang Y-M, Wong-Ng W (1982) J Am Chem Soc 104:5667
5. (a) Wiberg N, Wagner G (1983) Angew Chem Int Ed Engl 22:1005; (b) Wiberg N, Wagner G, Müller G (1985) Angew Chem Int Ed Engl 24:229
6. Apeloig Y, Bendikov M, Yuzefovich M, Nakash M, B-Zhivotovskii D (1996) J Am Chem Soc 118:12228
7. (a) Veszpremi T, Takahashi M, Hajgato B, Kira M (2001) J Am Chem Soc 123:6629; (b) Bendikov M, Solouki B, Auner N, Apeloig Y (2002) Organometallics 21:1349; (c) Grunenberg J (2001) Angew Chem Int Ed 40:4027; (d) Bendikov M, Quadt SR, Rabin O, Apeloig Y (2002) Organometallics 21:3930; (e) Gusel'nikov SL, Avakyan VC, Guselnikov SL (2002) J Am Chem Soc 124:662
8. Pannell KH (1970) J Organomet Chem 21:17
9. (a) Lewis C, Wrighton MS (1983) J Am Chem Soc 105:7768; (b) Randolph CL, Wrighton MS (1987) Organometallics 6:365
10. (a) Campion BK, Heyn RH, Tilley TD (1988) J Am Chem Soc 110:7558; (b) Campion BK, Heyn RH, Tilley TD (1992) J Chem Soc Chem Commun 1201; (c) Campion BK, Heyn RH, Tilley TD, Rheingold AL (1993) J Am Chem Soc 115:5527
11. Campion BK, Heyn RH, Tilley TD (1990) J Am Chem Soc 112:4079
12. Koloski TS, Carroll PJ, Berry DH (1990) J Am Chem Soc 112:6405
13. (a) Dioumaev VK, Plössl K, Carroll PJ, Berry DH (1999) J Am Chem Soc 121:8391; (b) Klei SR, Tilley D, Bergman RG (2001) Organometallics 20:3220
14. (a) Honda T, Satoh S, Mori M (1995) Organometallics 14:1548; (b) Honda T, Mori M (1996) Organometallics 15:5464; (c) Honda T, Mori M (1996) J Org Chem 61:1196; (d) Dekura F, Honda T, Mori M (1997) Chem Lett 825

15. (a) Mori M, Kuroda S, Dekura F (1999) J Am Chem Soc 121:5591; (b) Kuroda S, Sato Y, Mori M (2000) J Organomet Chem 611:304; (c) Kuroda S, Dekura F, Sato Y, Mori M (2001) J Am Chem Soc 123:4139
16. (a) Cardin DJ, Keppie SA, Kingston BM, Lappert MF (1967) Chem Commun 1035; (b) Muir KW (1971) J Chem Soc A 2663; (c) Kingston BM, Lappert MF (1972) J Chem Soc Dalton Trans 69
17. (a) Tilley TD (1985) Organometallics 4:1452; (b) Tilley TD (1985) J Am Chem Soc 107:4084; (c) Campion BK, Falk J, Tilley TD (1987) J Am Chem Soc 109:2049; (d) Elsner FH, Woo H-G, Tilley TD (1988) J Am Chem Soc 110:313; (e) Elsner FH, Tilley TD, Rheingold AL, Geib SJ (1988) J Organomet Chem 358:169; (f) Roddick DM, Heyn RH, Tilley TD (1989) Organometallics 8:324; (g) Arnold J, Engeler MP, Elsner FH, Heyn RH, Tilley TD (1989) Organometallics 8:2284; (h) Heyn RH, Tilley TD (1989) Inorg Chem 28:1768
18. Takahashi T, Hasegawa M, Suzuki N, Saburi M, Rousset CJ, Fanwick PE, Negishi E (1991) J Am Chem Soc 113:8564
19. Kreutzer KA, Fisher RA, Davis WM, Spaltenstein E, Buchwald SL (1991) Organometallics 10:4031
20. (a) Procopio LJ, Carroll PJ, Berry DH (1991) J Am Chem Soc 113:1870; (b) Procopio LJ, Carroll PJ, Berry DH (1993) Organometallics 12:3087
21. (a) Xue Z, Li L, Hoyt LK, Diminnie JB, Pollitte JL (1994) J Am Chem Soc 116:2169; (b) Wu Z, Diminnie JB, Xue Z (1998) Organometallics 17:2917; (c) Wu Z, McAlexander LH, Diminnie JB, Xue Z (1998) Organometallics 17:4853; (d) Wu Z, Diminnie JB, Xue Z (1999) Organometallics 18:1002
22. Campion BK, Heyn RH, Tilley TD (1993) Organometallics 12:2584
23. Negishi E, Cederbaum FE, Takahashi T (1986) Tetrahedron Lett 27:2829
24. Takahashi T, Swanson DR, Negishi E (1987) Chem Lett 623
25. Takahashi T, Suzuki N, Kageyama M, Nitto Y, Saburi M, Negishi E (1991) Chem Lett 1579
26. (a) Lappert MF, Raston CL, Engelhardt LM, White AH (1985) J Chem Soc Chem Commun 521; (b) Petersen JL, Egan JW Jr (1987) Organometallics 6:2007
27. Yoshifuji M, Loots MJ, Schwartz J (1977) Tetrahedron Lett 1303
28. (a) Lipshutz BH, Ellsworth EL (1990) J Am Chem Soc 112:7440; (b) Lipshutz BH (1997) Acc Chem Res 30:277
29. (a) Takahashi T, Kotora M, Kasai K, Suzuki N (1994) Tetrahedron Lett 35:5685; (b) Takahashi T, Hara R, Nishihara Y, Kotora M (1996) J Am Chem Soc 118:5154; (c) Takahashi T, Xi Z, Yamazaki A, Liu Y, Nakajima K, Kotora M (1998) J Am Chem Soc 120:1672; (d) Liu Y, Shen B, Kotora M, Takahashi T (1999) Angew Chem Int Ed 38:949

Octahedral Zirconium Complexes as Polymerization Catalysts

Anatoli Lisovskii · Moris S. Eisen (✉)

Department of Chemistry and Institute of Catalysis Science and Technology,
Technion- Israel Institute of Technology, 32000 Haifa, Israel
chmoris@tx.technion.ac.il

1	Introduction	64
2	Zirconium-Containing Octahedral Benzamidinate Complexes	65
2.1	Zirconium Bis(trimethylsilyl) Benzamidinate Complexes	66
2.2	N-Alkylated Benzamidinate Zirconium Complexes	73
2.3	Zirconium Mono-(N-trimethylsilyl)(N′-myrtanyl) Benzamidinates	76
2.4	Silica-Supported Bis(benzamidinate) Zirconium Dimethyl Complexes	80
3	Zirconium Allyl Complexes	87
4	Amido Zirconium Complexes	92
5	Zirconium-Containing Homoleptic Phosphinoamide Dynamic Complexes	97
6	The Structure of the Elastomeric Polypropylene	99
7	Conclusion	102
	References	102

Abstract The synthesis and X-ray structure of various octahedral zirconium complexes and their catalytic properties in the polymerization of α-olefins are described. Benzamidinate, amido, allylic, and phosphinoamide moieties comprise the study ligations. For the benzamidinate complexes, a comparison study between homogeneous and heterogeneous complexes is presented. For the phosphinoamide complex, we show that the dynamic symmetry change of the complex from C_2 to C_{2v} allows the formation of elastomeric polymers. By controlling the reaction conditions of the polymerization process, highly stereoregular, elastomeric, or atactic polypropylenes can be produced. The formation of the elastomeric polymers was found to be the result of the epimerization of the last inserted monomer to the polymer chain.

Keywords Octahedral complexes · Polymerization · Elastomers · Propylene · Catalysis

1
Introduction

Since the pioneering work of Ziegler and Natta applying the catalytic system TiCl$_4$/AlClEt$_2$ to the polymerization of ethylene to high-density polyethylene, and propylene to stereoregular polypropylene [1-3], new metal-mediated catalysts for the polymerization of olefins have experienced a wide-ranging development. These studies have resulted in the discovery of metallocenes and geometry-constrained complexes as new generations of Ziegler–Natta catalysts for the polymerization of α-olefins [4-9]. Being activated with methylalumoxane (MAO) or perfluorinated boron cocatalysts, the metallocenes show the highest polymerization activities, producing polymers with high stereospecificity as compared with conventional Ziegler–Natta catalysts. Having a single-site nature, the metallocenes produce polymers with high compositional uniformity and narrow polydispersities, allowing sophisticated control over the polymer architecture [4-12].

In general, the active species in the polymerization of olefins with metallocenes comprise the cyclopentadienyl ancillary ligations (L$_n$), an electron-deficient metal center (M), and a weakly coordinative counteranion (cocatalyst) (X$^-$)

R = alkyl group

In addition to metallocenes and geometry-constrained complexes, over the last 20 years a tremendous effort has been made in designing new complexes not containing cyclopentadienyl (Cp) moieties, as potential Ziegler–Natta catalysts for the polymerization and copolymerization of α-olefins [13, 14]. Various compounds of the Group 3–13 metals with carbon- [15-17], oxygen- [18-21], and nitrogen- [22-26] based ligands have been described.

Previously, we synthesized and studied various Group 4 complexes with different ligations as alternatives to the cyclopentadienyl ligand. Here we present an overview of the synthesis, structure, and catalytic properties in the polymerization of α-olefins of several zirconium octahedral complexes. We show how the stereoregular polymerization of α-olefins using these octahedral zirconium complexes can be modulated by pressure. These results raise conceptual questions regarding the general applicability of *cis*-octahedral C$_2$-symmetry complexes to the stereospecific polymerization of α-olefins.

2
Zirconium-Containing Octahedral Benzamidinate Complexes

Within the variety of the designed ligands, bulky heteroallylic compounds have been described. Such complexes usually contain the benzamidinate fragment in a chelating coordination mode.

In most cases the substituents R and R' are alkyl, aryl, or SiMe$_3$. The most thoroughly investigated benzamidinates are the Zr- and Ti-containing complexes with N-TMS (TMS=trimethylsilyl) substituents. These systems, considered as steric equivalents of Cp or Cp* (Cp=C$_5$H$_5$, Cp*=C$_5$Me$_5$), are unique in their electronic properties [27–30]. The anionic moiety [R-C(NSiMe$_3$)$_2$]$^-$, being a four-electron donor, promotes higher electrophilicity at the metal center compared to the six electrons of the Cp ligands. The possibility of modifying both the steric bulk and electronic properties of the benzamidinates through changes in their composition and structure, made these ligands very attractive for the synthesis of various organometallic complexes, widely used in the polymerization of α-olefins as analogs of the metallocenes.

The polymerization of α-olefins using Group 4 chelating benzamidinate complexes as precatalysts has been investigated by various research groups including ours [31–40]. These complexes are normally obtained as a mixture of racemic C$_2$-symmetry structures, in which the central metal is octahedrally coordinated by the two chelating benzamidinate ligands and two chlorine atoms or methyl groups. When activated with MAO or other cocatalysts, these complexes were found to be active catalytic precursors for polymerization and other olefin transformations.

Later on, as a development of our earlier investigations, a systematic study was conducted of the synthesis of various Group 4 chelating benzamidinate complexes and their applications as polymerization catalysts [31,41]. The goal of these investigations was to examine the relationship between the structural and compositional features of the Group 4 chelating benzamidinate complexes, and the microstructure and stereoregularity of the polymers obtained. In addition, different effects in the polymerization process (temperature, pressure, nature and amount of the cocatalyst, and solvent) were studied, and a plausible mechanism for the polymerization of propylene promoted by racemic mixtures of cis-octahedral complexes was proposed. These complexes are considered to be new catalytic precursors for the highly stereoregular polymerization of propylene, which can be modulated by pressure. It has been shown that in the polymerization of propylene at atmospheric pressure, an oily product resembling an atactic polymer was produced, rather than the expected isotactic polypropylene [33]. However, performing the reaction at high pressure (in liquid propylene) resulted in the formation of a highly stereoregular polymer. Thus, it was demonstrated that the stereoregular polymerization of propylene catalyzed by *cis*-octahedral benzamidinate early transition metal complexes with C_2 symmetry, activated with either MAO or $B(C_6F_5)_3$ cocatalysts, can be tailored by pressure (from atactic to isotactic through elastomers).

2.1
Zirconium Bis(trimethylsilyl) Benzamidinate Complexes

The N,N'-bis(trimethylsilyl)benzamidinate zirconium dichloride complexes $[\eta^2\text{-}4\text{-}RC_6H_4C(NSiMe_3)_2]_2ZrCl_2$ (R=H (**1**) or CH_3 (**2**)) were prepared by the reaction of $ZrCl_4$ with two equivalents of bis(trimethylsilyl)benzamidinate lithium·TMEDA (TMEDA=N,N,N',N'-tetramethylenediamine) at room temperature in toluene yielding, after removal of the TMEDA and crystallization, 83–89% of pale yellow analytically pure rhombohedral crystals of **1** and **2** (Scheme 1) [31, 33]. These complexes were also obtained in a high quality form

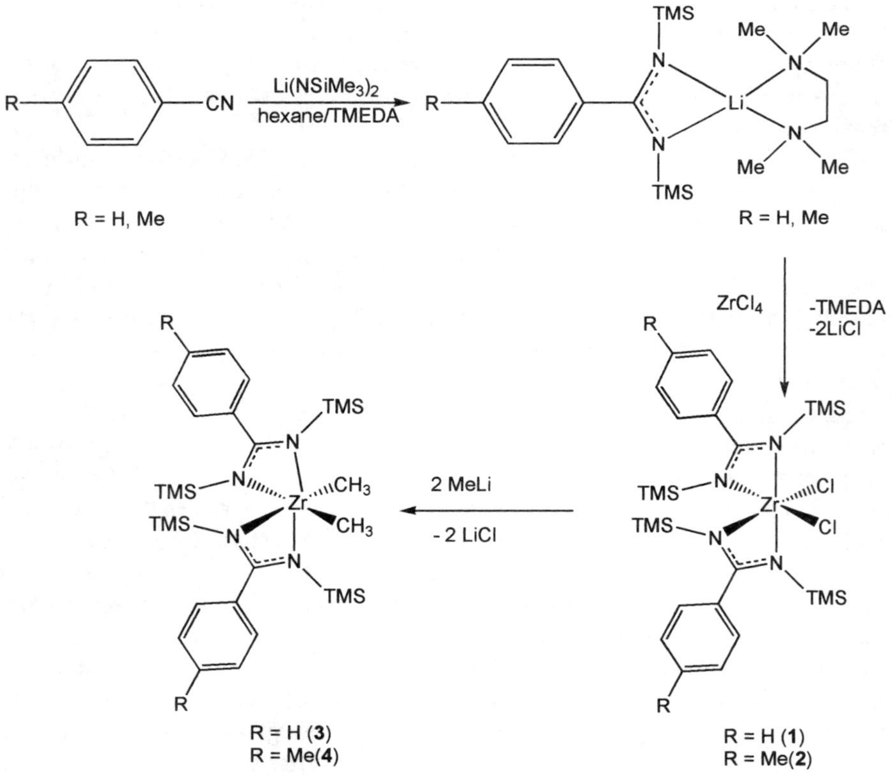

Scheme 1 Synthesis route to the bis(benzamidinate) zirconium complexes **1–4** [41]

Octahedral Zirconium Complexes as Polymerization Catalysts

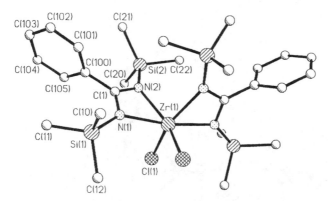

Fig. 1 Molecular structure of bis(benzamidinate) zirconium dichloride (1) [31]

in terms of purity and crystallinity, although in lower yields (55%), by the reaction of $ZrCl_4$ and one equivalent of the benzamidinate Li dimer [42]. Alkylation of complexes 1 and 2 with two equivalents of MeLi·LiBr yields yellow dimethyl complexes 3 and 4, respectively (Scheme 1).

The molecular structure of the benzamidinate zirconium dichloride (complex 1) consists of a Zr atom chelated by two benzamidinate ligands and two chlorine atoms to give a distorted octahedral environment at the metal (Fig. 1) [31]. The benzamidinate ligands form four-membered rings almost coplanar with the Zr atom (torsion angle Zr–N1–C1–N2=1.4°). The Zr–Cl distance (2.401(2) Å) and Zr–N distances (Zr–N1=2.238(5), Zr–N2=2.204(5) Å) are significantly elongated with respect to those of the analogous titanium complex [31].

It is noteworthy that the benzamidinate ligands lie in two nonsymmetrical planes. Comparing the benzamidinate ligation in these complexes with metallocenes, using the benzamidinate central carbon as the centroid, shows that between two Zr atoms the cone angle is 124.8(3)°, whereas in metallocenes this angle is 136°.

The molecular structure of the dimethyl complex 4 is shown in Fig. 2 [33, 41]. Here the central Zr atom is octahedrally bonded to two benzamidinate ligands and two terminal methyl groups. One carbon atom from the CH_3 groups (C1) and one nitrogen atom from the chelating benzamidinate unit are in the axial positions, producing a C_2-symmetry complex. The distances Zr–N1 to Zr–N4 in complex 4 are somewhat longer than those in the dimeric compound $[C_6H_4C(NSiMe_3)_2ZrCl_3]_2$ (2.14 and 2.19 Å) [43], while they are similar to those in complex 1 or other zirconium-containing monomeric complexes [37, 44]. As well as for complex 1, the benzamidinate moieties in 4 are located almost in one plane with the corresponding angles Zr–C3–C4 and Zr–C11–C12 of 168.7(3) and 170.6(4)°, respectively.

Complexes 1 and 2 were studied as catalytic precursors in the polymerization of ethylene and the results are summarized in Table 1 [31]. Raising the temperature from 5 to 60 °C causes an increase in the catalytic activity by a

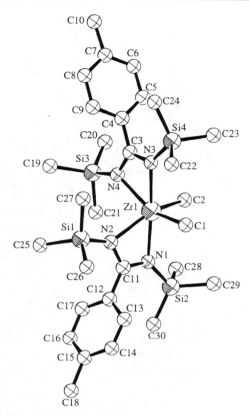

Fig. 2 ORTEP diagram of complex 4 [41]

Table 1 Data for the polymerization of ethylene catalyzed by complexes 1 and 2[a] [31]

Entry	Catalyst[b]	Al:Zr ratio[c]	T (°C)	Activity[d] (×10⁻⁴)	M_v^e	m.p. (°C)
1	1	200	25	3.2	150,000	133.9
2	1	400	5	0.65	48,000	124.8
3	1	400	25	2.6	50,000	129.6
4	1	400	60	2.8	78,000	132.9
5	2	400	25	1.6	30,000	130.5
6[f]	1	400	25	57	162,000	135.4

[a] 1 atm, in toluene.
[b] 2.9×10⁻⁴ mol.
[c] MAO obtained from a 20% solution in toluene after evaporation of solvent in vacuum at 25 °C.
[d] g polymer·mol Zr⁻¹·h⁻¹.
[e] By viscosimeter technique in 1,2,4- trichlorobenzene at 130 °C (K=5.96×10⁻⁴, a=0.7).
[f] 5 atm.

factor of ca. 4 (entries 2 and 4). It can be seen that the smaller the Al:Zr ratio, the higher the activity and molecular weight of the polymer (entries 1 and 3).

These results, opposite to those observed for the cyclopentadienyl early transition metal systems, can be accounted for in terms of the known influence of the cocatalyst concentration on the possible eliminations, alkyl transfer pathways, and other deactivation processes [45, 46]. Under similar conditions, the polymerization of ethylene at high pressure leads to a considerable increase in activity and produces polymers of higher molecular weight than at atmospheric pressure (entries 6 and 3). This effect is a consequence of the rate of insertion, which is proportional to the monomer concentration in solution.

The electronic effect of the benzamidinate ligand is clearly shown by comparing the activities of complexes 1 and 2. The former, containing a hydrogen atom in the *para* position of the aromatic ring, is more active in the polymerization of ethylene than complex 2, in which the hydrogen atom was substituted by the CH_3 group (entries 3 and 5). A plausible elucidation relates to the difference in the electronic effects of both ligations on the cationic Zr center. The methyl group, with a more negative inductive effect, that is, greater electron–donor properties than the hydrogen atom [47], increases the electron density at the phenyl ring, thus decreasing the positive charge at the metal center. The reduction in the cationic character of the latter induces a weaker bond with an incoming molecule of olefin, reducing the polymerization activity of the complex.

Complex $[p\text{-MeC}_6H_4C(NSiMe_3)_2]_2ZrMe_2$ (4) was found to be a good precursor for the polymerization of propylene at atmospheric pressure. Being activated with MAO in toluene (Al:Zr ratio=250:1), this complex produces large amounts (8.1×10^5 g polymer·mol $Zr^{-1} \cdot h^{-1}$) of an oily polymer with molecular weight Mw=38,000, resembling an atactic polypropylene. ^{13}C-NMR triad and pentad analysis showed that, contrary to the ten expected signals for the CH_3 groups of an atactic polypropylene, only two major signals were observed in the region between 19.5 and 21.9 ppm (Fig. 3) [48]. (A similar spectrum has recently been obtained for the polypropylene polymerized by α-diimine nickel complexes [49]). These two signals belong to the corresponding *mr* and *rr* triads. The small signals in the region of 15 to 19 ppm and the signals between 29 and 41 ppm corroborate the presence of 2,1 and 1,3 misinsertions. Moreover, it is possible to conclude that β-methyl elimination is the major termination mechanism, as represented by the signals at 146 and 113 ppm, and the signals between 22 and 26 ppm for the vinylic CH (C1), CH_2 (C2), and the terminal carbons of the isobutylene unit (C3, C4, and C5) in Fig. 3, respectively [48].

When 4 was activated with $B(C_6F_5)_3$ (molar ratio B:Zr=1), under the same reaction conditions as with MAO, a highly isotactic polypropylene (*mmmm*= 98%) was formed, contrary to the results obtained with MAO (Eq. 1). The activity of this catalytic system (1.2×10^5 g polymer·mol $Zr^{-1} \cdot h^{-1}$) is lower than the activity of the complex 4 activated with MAO. It is important to point out that complex 4 has a C_2-symmetry octahedral geometry, suggesting that theoretically, when activated with MAO, an isotactic polypropylene should also be expected, as it was in the case of the boron-containing cocatalyst.

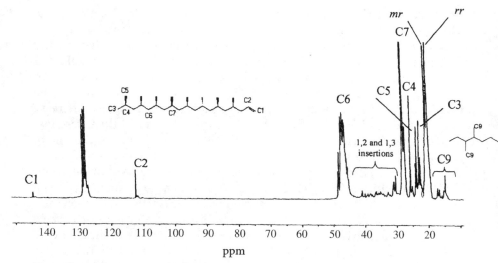

Fig. 3 ^{13}C-NMR spectrum of the polypropylene obtained with complex 4 [41]

The "atactic polymer" obtained with MAO can be rationalized by an intramolecular epimerization reaction of the growing polypropylene chain at the last inserted monomeric unit [50–57], which at low monomer concentration was found to be faster than the stereoregular insertion of propylene (Scheme 2a).

When the reaction catalyzed with the system "complex 4/MAO" is carried out at higher pressure, a highly stereoregular polymer is formed (Table 2). By performing the polymerization in dichloromethane, larger activities, stereoregularities, and melting points of the polymers were achieved, as compared to the results obtained in toluene (entries 1 and 2). This increase in activity is plausibly a consequence of the polarity of the CH_2Cl_2, causing a greater charge separation between the cationic benzamidinate alkyl complex and the MAO anion, encouraging the insertion of the monomer, as compared to toluene [58, 59]. For the latter, a putative π-bonding of the ring to the cationic center

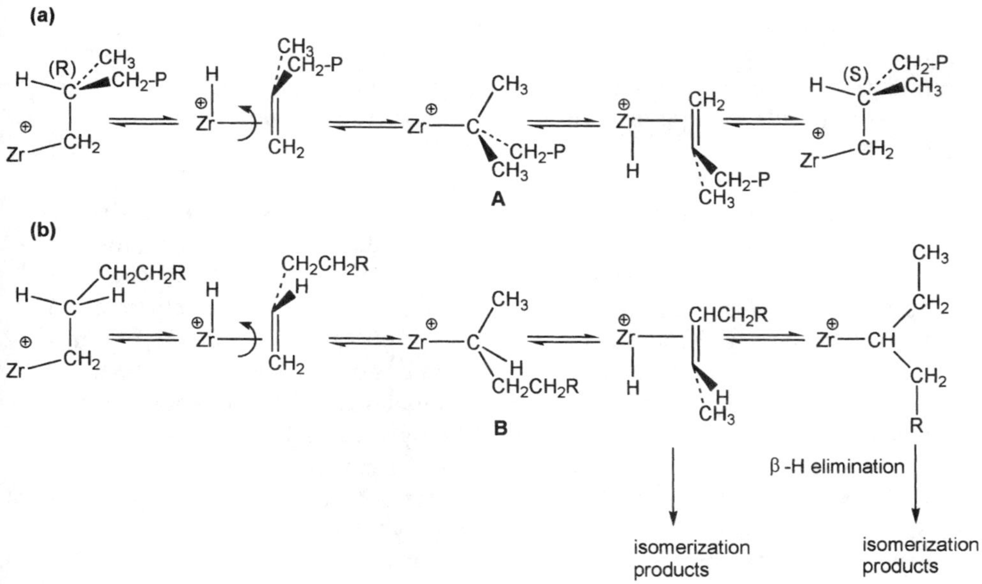

Octahedral Zirconium Complexes as Polymerization Catalysts 71

Scheme 2a, b Proposed mechanism for the intramolecular epimerization of the growing polypropylene chain (a), and plausible mechanism for the expected isomerization of α-olefins (b)

Table 2 Results for the polymerization of propylene with complex 4 at high pressure[a] [41]

Entry	Solvent	Al:Zr	P (atm)	T (°C)	Activity[b] ($\times 10^{-5}$)	mmmm[c] (%)	Mw	mwd[d]	m.p. (°C)
1	Toluene	250	9.2	25	1.1	86[e]	440,000	1.69	142
						11	36,000	2.35	Oil
2	CH_2Cl_2	250	9.2	25	2.2	90	27,000	2.49	146
3	CH_2Cl_2	400	9.2	25	7.5	96	42,000	1.81	147
4	CH_2Cl_2	1000	9.2	25	7.9	98	83,000	1.42	149
5	CH_2Cl_2	250	5.1	0	0.5	86	19,000	1.85	138
6	CH_2Cl_2	250	9.2	25	2.2	90	27,000	2.49	146
7	CH_2Cl_2	250	17.0	50	26.6	98	271,000	1.81	152
8	Toluene	$B(C_6F_5)_3$	9.2	25	2.8[f]	98	51,000	1.96	154
						7	9,300	3.10	Oil

[a] 7.3×10^{-6} mol of 4.
[b] g polymer·mol $Zr^{-1} \cdot h^{-1}$.
[c] Measured by ^{13}C-NMR.
[d] Molecular weight distribution.
[e] Isotacticity of the isotactic fraction.
[f] 70% atactic and 30% isotactic.

of the complex possibly takes place forming a cationic η^6-toluene compound [60–63]. This intermediate inhibits the insertion of the monomer and the termination of the polymer chain, allowing larger molecular weights and a narrower molecular weight distribution as compared to the polymers formed in CH_2Cl_2.

In dichloromethane, the polymerization of propylene under higher pressure yields an isotactic polypropylene with a small number of stereodefects, while in toluene a mixture of isotactic and atactic products was obtained. This fact can testify about the presence of two different active catalytic species. The nature of the intermediate responsible for the formation of the atactic fraction has not been fully elucidated yet. On the basis of NMR experiments, it seems that in toluene one of the benzamidinate ligations opens to an η^1 coordination, and the solvent occupies the vacant site. It was shown that similar heteroallylic ligations (alkoxysilyl-imido) have exhibited dynamic behaviors [64, 65].

Up to a molar ratio of Al:Zr=400, the activity of the catalyst is enhanced with an increase in the MAO amount (entries 2–4). Further augmentation of MAO concentration does not influence the polymerization rate, but rather the properties of the polypropylene. By raising the Al:Zr ratio, the melting points, molecular weights, and isotacticities of the polymers increase with a concomitant reduction in their polydispersities. When the reaction was carried out at elevated temperatures, higher activities, stereoregularities, and molecular weights of polymer were achieved (entries 5–7). It may be suggested that an increase of the temperature induces faster insertion rates as compared to the rate of the epimerization, resulting in higher isotacticities.

In CH_2Cl_2, the polymerization of propylene with 4 activated by $B(C_6F_5)_3$ does not take place, while in toluene (entry 8) two polymeric fractions (70% atactic and 30% highly isotactic) were attained, as observed with MAO (entry 1).

Comparing the stereoregularities of the polymers obtained with the catalytic systems 4/MAO and 4/$B(C_6F_5)_3$ at atmospheric and higher pressure, a remarkable effect of the counterion is observed. When the boron-containing cocatalyst is used at atmospheric pressure in toluene, a highly isotactic polymer is obtained, while at higher propylene concentration two polymer fractions are accountable. The opposite effect is observed for MAO, which allows large isotacticities at higher pressure but only in dichloromethane. Thus, the stereoregularity of polypropylene obtained with the system 4/MAO can be tailored as shown in Eq. 2.

^{13}C-NMR analysis of the chain ends of the polymers formed at higher pressure also shows that the β-methyl elimination is the exclusive termination chain mechanism for these catalytic systems. Statistically, it was deduced that at atmospheric pressure, the epimerization is observed in every small number of insertions, forming a *-mmrrmmrrmmrrmmrr-* microstructure responsible for the two major signals with similar integration, and the almost absent signal for the *mmmr* pentad, as presented in Fig. 3. The lack of 2,1 and 1,3 misinsertion signals in the ^{13}C-NMR spectra of the polymers obtained at higher pressure indicates the rapidity of the 1,2 insertion, as compared to the misinsertions.

Additional endorsement indicating that the mechanism shown in Scheme 2a is responsible for the stereodefects in polypropylene produced by the catalytic system 4/MAO was obtained by reacting this system with 1-octene [66]. Because the latter polymerizes at an extremely low rate, the epimerization reaction, if applicable, should induce the isomerization of α-olefins to the corresponding internal olefins. A β-hydrogen elimination is expected to occur either from the β-methyl group in complex **B** (Scheme 2b), causing no change in the 1-octene, or from the α-position at the alkyl group (CH$_2$–CH$_2$–R) attached to the polymer chain (**B** in Scheme 2b). This second pathway induces the isomerization of the double bond. The fact that the complex [p-MeC$_6$H$_4$C(SiMe$_3$)$_2$]$_2$ZrMe$_2$ activated with MAO in toluene-d_8 (Al:olefin:catalyst molar ratio=400:180:1, 10 mg of **4**, 0.6 ml toluene, 85 °C, 6 h) catalyzes the isomerization of 1-octene yielding 2-(E)-octene (25%), 2-(Z)-octene (15%), 3-(E)-octene (40%), and *trans*-4-(E)-octene (18%), displayed strong evidence for the formation of this intermediate.

2.2
N-Alkylated Benzamidinate Zirconium Complexes

The synthesis of Zr-containing N-alkylated benzamidinate complexes and their use in the polymerization of α-olefins have been also studied [32]. The bis(dialkylbenzamidinate) zirconium dichloride [C$_6$H$_5$C(NC$_3$H$_7$)$_2$]$_2$ZrCl$_2$ was prepared by reaction of ZrCl$_4$(THF)$_2$ with two equivalents of Li[C$_6$H$_5$C(NC$_3$H$_7$)$_2$], affording the disubstituted complex 5 as a yellow crystalline solid (Eq. 3).

Complex 5 (Fig. 4) is stable in air and less sensitive to hydrolysis than the corresponding N-trimethylsilyl derivatives. The central Zr atom is octahedrally surrounded by the two chelating benzamidinate ligands and two terminal chlorine atoms. One chlorine atom (Cl1a) and one nitrogen atom of a chelate unit (N2a) are in the axial positions, while the second chlorine atom and the remaining N atom occupy the equatorial positions. Therefore the C–N bond lengths of the two chelate units are slightly different (1.32 and 1.35 Å). The identical Zr–N distances (2.21 and 2.22 Å) are alike in the benzamidinate zirconium dichloride [31]. Remarkable is the large torsion angle between the two planar ZrCN$_2$ rings, which are almost perpendicular with respect to each other

$$\text{ZrCl}_4(\text{THF})_2 + 2\text{Li}[\text{C}_6\text{H}_5\text{C}(\text{N}(i\text{-Pr}))_2] \xrightarrow[-2\text{LiCl}]{\text{THF}}$$

(3)

(5)

(89.2°). This obviously minimizes repulsion between the bulky isopropyl groups. The torsion angles between the phenyl rings and the ZrCN$_2$ planes are 89.3°, which is the highest value for any metal benzamidinate complex.

The results of the polymerization of α-olefins with complex 5 activated by methylalumoxane are given in Table 3. Comparison of the data of Tables 1 and 3 testifies that as well as for the catalyst [C$_6$H$_5$C(NSiMe$_3$)$_2$]$_2$ZrCl$_2$ (complex 1), raising the temperature in the ethylene polymerization with [C$_6$H$_5$C(NC$_3$H$_7$)$_2$]$_2$ZrCl$_2$ (complex 5) markedly increases the reaction rate (Table 3, entries 1, 4, and 7). However, contrary to 1, the melting point of the polymers formed with 5 slightly decreases with increasing polymerization temperature. For both complexes, the opposite regularity is also observed with regard to the Al:Zr ratio effect; as distinct from complex 1, for 5 increasing the MAO content results in an increase the catalytic activity (entries 3–5 for 25 °C, and 6–8 for 60 °C), reaching a plateau limit with larger excess of the cocatalyst.

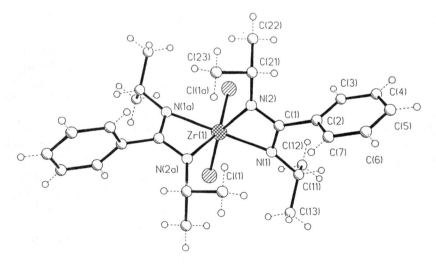

Fig. 4 The molecular structure of complex 5 [32]

Octahedral Zirconium Complexes as Polymerization Catalysts 75

Table 3 Data for the ethylene polymerization catalyzed by complex 5[a] [32]

Entry	Al:Zr ratio	T (°C)	Activity[b] (×10⁻⁴)	m.p. (°C)
1	800	0	1.4	135.7
2	200	25	0.68	136.6
3	400	25	3.4	133.3
4	800	25	5.6	133.4
5	2,000	25	6.0	134.4
6	400	60	8.6	130.9
7	800	60	21.0	131.8
8	2,000	60	40.0	133.0

[a] 1 atm, 4.4×10⁻⁴ mol of catalyst, solvent toluene, reaction time 5 min, MAO removed from a 20% solution in toluene at 25 °C/10⁻⁶ Torr.
[b] g polymer·mol Zr⁻¹·h⁻¹.

Polymerization of propylene with complex 5 (Table 4) at atmospheric pressure produces an "atactic" polypropylene having the same features regarding the temperature and Al:Zr ratio as for ethylene. The ^{1}H and ^{13}C-NMR spectroscopic analysis of polypropylene reveals only vinyl/isopropyl, but no vinylidene/n-propyl, end groups, similar to the polymers obtained with zirconocenes [67, 68]. Polymers with these end groups may be formed from at least three different mechanisms. The first involves an allylic C–H activation of propylene, the second, a β-methyl elimination, and the third, a β-hydrogen elimination from a polymer chain in which the monomer inserts in a 2,1 fashion [69]. Since in the

Table 4 Results for the polymerization of propylene catalyzed by complex 5[a] [32]

Entry	Solvent	Al:Zr ratio	T (°C)	P (atm)	Activity[b] (×10⁻⁴)	M_v	m.p. (°C)
1[c]	Toluene	400	25	1.0	0.68	–[d]	–[d]
2[c]	Toluene	1,000	25	1.0	1.4	–[d]	–[d]
3[c]	Toluene	4,000	25	1.0	2.4	–[d]	–[d]
4[c]	Toluene	1,000	60	1.0	2.1	–[d]	–[d]
5	CH$_2$Cl$_2$	200	25	7.2	0.27	87,000	145.7
6	CH$_2$Cl$_2$	400	25	7.2	0.47	29,000	150.5
7	CH$_2$Cl$_2$	400	50	10.1	0.11	20,000	145.8
8	Toluene	400	25	7.2	0.7	42,000	148.7

[a] 4.4×10⁻⁴ mol of 5, reaction time 5 min.
[b] g polymer·mol Zr⁻¹·h⁻¹.
[c] An atactic polypropylene is formed.
[d] The amount of polymer formed during a few minutes of the reaction was not enough for the measurements.

polymer obtained no *n*-propyl end groups were observed, the last mechanism can be rejected. As the time of the polymerization was up to 5 min and larger duration of the reaction did not yield more polymer, it can be suggested that besides the β-methyl elimination pathway, which presumably is the major chain termination process, a competitive deactivation pathway inhibits polymer formation.

Polymerization of propylene with 5 at higher pressure results in the formation of isotactic polymer in both solvents (dichloromethane and toluene). In CH_2Cl_2, increasing the MAO concentration leads to an increase of the polymerization rate and melting point of the polymer, although the molecular weight of the latter decreases (entries 5 and 6). Raising the temperature results in decreases of the polymer yield, its molecular weight, and melting point (entries 6 and 7). Under the same conditions, polymerization activity of the 5/MAO system in toluene is higher than that in dichloromethane, and the polymer formed has a higher molecular weight (compare entries 6 and 8).

It is important to point out that the zirconium benzamidinate complexes 1 and 5 have lower catalytic activity in the α-olefin polymerizations than the corresponding metallocene systems (1×10^7 and 6×10^6 g polymer·mol $Zr^{-1} \cdot h^{-1}$ for ethylene and propylene, respectively [70]). The difference in activity can be rationalized by the structural environment of the metal (Zr) center. In the metallocenes, the cone angle is normally the key feature for unsaturation, whereas in the octahedral complexes the groups at the N-moiety may induce an electronic unsaturation at the metal center. Thus, the lower polymerization activities of the complexes 1 and 5 may be a consequence of the partial saturation at the metal orbitals as compared to the metallocenes [71–73].

2.3
Zirconium Mono-(*N*-trimethylsilyl)(*N'*-myrtanyl) Benzamidinates

The first example of the synthesis and structural characterization of the chiral Zr complexes C_3-tris[(*N*-SiMe$_3$)(*N'*-myrtanyl) benzamidinate]$_3$ZrR (R=Cl (6), R=Me (7)), and their use as precursors for the polymerization of propylene, has been recently described [74]. The chiral benzamidinate ligand was prepared from the corresponding trimethylsilylmyrtanylamine after lithiation with BuLi and the concomitant reaction with benzonitrile. Reaction of three equivalents of the ligand with $ZrCl_4$ in toluene or THF at room temperature yielded 25–30% of complex 6. After alkylation of the latter with an excess of MeLi·LiBr in toluene at room temperature, a brown solution of a mixture of complexes 6 and 7 was obtained (Scheme 3). Crystallization of the mixture afforded single cocrystals containing 70:30 of 6 and 7, respectively.

X-ray study of 6 revealed a "propeller"-like structure (Fig. 5), in which three SiMe$_3$ groups are in a *cis* position with respect to the chlorine atom. The other three myrtanyl groups are arranged on the opposite side of the chlorine. The distances between the benzamidinate moiety or the chloride atom and the Zr center are normal as compared to nonchiral Group 4 and actinide benzamid-

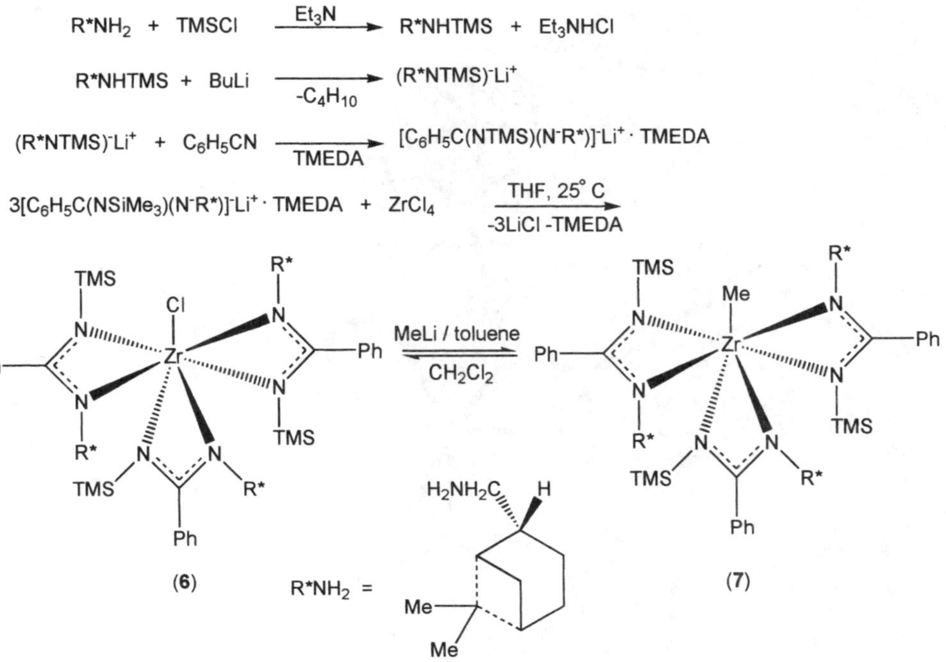

Octahedral Zirconium Complexes as Polymerization Catalysts 77

Scheme 3 Synthetic procedure for the preparation of complexes 6 and 7

inate complexes (Zr–N=2.24–2.35 Å, Zr–Cl=2.46–2.48 Å) [35, 75]. The X-ray structure of the cocrystalline complex is similar to that of 6 with a Zr–Me distance of 2.469(10) Å [35].

Comparison of the structures of complex 6 and the nonchiral compound [$C_6H_5C(NSiMe_3)_2$]$_3$ZrCl shows that both complexes crystallize in the hexagonal crystal system having P3 and Pc1 space groups, respectively [35]. Both complexes have a capped octahedral propeller-like structure, but in the nonchiral complex both enantiomeric propeller structures are present in the crystal structure, whereas in the chiral complex only one enantiomeric structure is observed. The major distinction between the two complexes is the position of the three benzamidinate ligands. Thus, in 6 the blades in the propeller are pushed down opposite the chloride ligand as compared to the nonchiral complex. The angle between the propeller flap and the chlorine atom is larger by 8° as related to the racemic complex, exhibiting a more exposed chloride ligand. This opened coordination is likely responsible for the different polymerization activities of both complexes.

When activated with MAO complex 6, under pressure, polymerizes propylene producing isotactic polymer (Table 5). The fact that in the ^{13}C-NMR spectrum of the polymers no *mrmm* signals were found, can testify about the formation of the polymers exclusively by the site control mechanism. Increasing the temperature induces an increase of the reaction rate, but the molecular weight

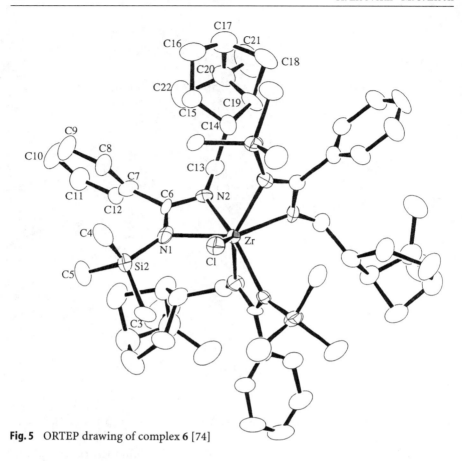

Fig. 5 ORTEP drawing of complex 6 [74]

Table 5 Results of the propylene polymerization with complex 6 activated by MAO[a] [74]

Entry	Solvent	Al:Zr ratio[b]	T (°C)	P (atm)	Activity[c] (×10⁻³)	mmmm (%)	M_v[d]	m.p. (°C)
1	Toluene	280	80	12.0	9.23	79.5	7750	128.5
2	Toluene	280	50	8.6	0.716	92.8	10,300	135.0
3	Toluene	280	0	5.1	0.200	98.9	32,000	145.0
4	Toluene	560	50	8.6	1.06	95.0	33,700	140.0
5	Toluene	830	50	8.6	0.105	93.8	250	139.4
6	CH$_2$Cl$_2$	280	50	8.6	5.29	–	930[e]	–

[a] 4.5×10⁻⁶ mol of 6; 20 ml solvent.
[b] MAO obtained from 20% solution in toluene (Schering) by vacuum evaporation of solvent at 25 °C/10⁻⁶ Torr.
[c] g polymer·mol Zr⁻¹·h⁻¹.
[d] By viscosimeter technique in 1,2,4-trichlorobenzene at 130 °C (K=1.37×10⁻², a=0.7).
[e] Atactic polypropylene.

of the polymers decreases. Using the polar solvent (CH_2Cl_2) leads to a decrease in activity and to formation of low molecular weight oily polypropylene, probably due to the formation of the dichloro complex again. Increasing the MAO concentration results in an increase in the polymerization rate and molecular weight of the polymers, reaching a maximum in activity and molecular weight at an Al:Zr ratio of about 600. In the polymerization of propylene with the catalytic system 6/MAO at atmospheric pressure, polypropylene is not formed [74].

We have shown that bis(benzamidinate) zirconium dichloride (complex 2), activated with MAO, at high pressure polymerizes propylene yielding a highly stereoregular polypropylene, as well as the system 6/MAO. However, contrary to the latter, at atmospheric pressure the system 2/MAO produces, at a high rate, an oily polymer resembling an atactic polypropylene. The different behavior of the complexes 2 and 6 at atmospheric pressure can be plausibly explained as follows. In the case of 6, at high pressure one ancillary ligand is not acting as a spectator ligand, and MAO is able to extract one amido–imine moiety, producing the chiral cationic complex [7, 81]. Taking into account that a high propylene concentration is required and that aluminum benzamidinate complexes are known [82, 83], we expect that the displacement of the benzamidinate ligand should proceed most likely through an $\eta^3 \rightarrow \eta^1$ slippage mechanism (Eq. 4). Although two different η^1-amido configurations are possible theoreti-

(4)

cally, the preferred configuration will contain a SiMe$_3$ group at the β position to the metal center [76-78]. We have shown that metallocene complexes with amido, phosphido, and arsenido ligations bearing a SiMe$_3$ group at the β position are easily cleaved with MAO, allowing the formation of the active cationic metallocenes for the polymerization of propylene [79, 80].

2.4
Silica-Supported Bis(benzamidinate) Zirconium Dimethyl Complexes

For achieving high catalytic activity, metallocenes and non-Cp organometallic complexes require a large excess of the expensive MAO cocatalyst. Therefore, many studies have been conducted to replace methylalumoxane by other cocatalysts. For example, using perfluoroaromatic cocatalysts, only stoichiometric amounts are required [10, 84-86]. Other disadvantages generally inherent in using homogeneous catalysts for the polymerization technology are reactor fouling, the necessity to separate the polymers from the catalytic mixture, and the recovery of the solvents. These problems, in principle, can be overcome by the heterogenization of the homogeneous catalysts by means of their immobilization on porous supports.

The most used solids for the heterogenization of metallocenes of early transition metals are silica and zeolites [87-90]. Supported metallocenes generally show less activity as compared to the corresponding homogeneous systems, but the anchored complexes have an important advantage: the amount of MAO can be markedly reduced or even replaced with AlR$_3$ [87-89]. Besides, in the heterogeneous systems, the active catalytic species are stabilized inducing an easier control of the polymer morphology and bulk density [91].

As distinct from metallocenes, supported non-Cp organometallic complexes are less studied, but the number of research projects devoted to this topic is markedly increasing. For example, various "constrained-geometry" titanium complexes were immobilized on different supports and studied in the polymerization of ethylene, propylene, and styrene [89]. These supported complexes allow the formation of high molecular weight polyethylene, polypropylene with elastomeric properties, and syndiotactic polystyrene. Recently it was shown that the chelating diamide dimethyl titanium complex ([ArN(CH$_2$)$_3$NAr]TiMe$_2$, Ar=2,6-^{i}Pr$_2$C$_6$H$_3$), immobilized on the MMAO/SiO$_2$ support (MMAO=modified methylalumoxane) catalyzed the living polymerization of propylene at 0 °C [92].

We have shown that zirconium benzamidinate complexes produced either highly isotactic polymer (high pressure) or atactic polypropylene (atmospheric pressure). For the corresponding titanium complex, at high propylene concentration an elastomeric polymer was formed. Since the number of stereodefects affecting the properties of the polypropylene presumably depends on the epimerization rate of the last inserted molecule of monomer, it was of interest to study the effect of supporting these complexes on mesoporous solids. The basic idea consisted in assuming that monomer approaching the ac-

Fig. 6 Structure of MCM-41 showing the honeycomb formation of the silica matrix

Fig. 7 Wormlike structure of hexagonal mesoporous silica (HMS)

tive catalytic center will depend on the structural features of the support, thus inducing formation of an elastic or an atactic polymer, respectively. The propylene polymerization was performed with the $[p\text{-MeC}_6\text{H}_4\text{C(NSiMe}_3)_2]_2\text{ZrMe}_2$ (complex 4) supported on mesoporous silica (MCM-41 and HMS) [93–101].

The choice of the inorganic supports was based on their different structural features. The MCM-41 material consists of uniform hexagonal arrays of linear channels that are constructed with a silica matrix like a honeycomb (Fig. 6), whereas HMS (hexagonal molecular silica) has a wormlike or sponge structure (Fig. 7) [94–100].

Samples of the supports were prepared according to described literature procedures [93, 94, 99–101]. Both MCM-41 and HMS silicas were preliminarily dehydrated at 400 °C at high vacuum. The supports were impregnated with MAO in toluene and, after washing with toluene and drying in vacuum at room temperature, the MAO/support solids were stirred with a certain amount of complex 4 in toluene at 50 °C for 3 h (Eq. 5). The solids were filtered, washed with toluene, and dried under vacuum at room temperature to obtain the supported catalysts 4/MAO/MCM-41 and 4/MAO/HMS.

The results of the polymerization of α-olefins using both heterogeneous complexes are given in Table 6. For comparison, the data obtained with the corresponding homogeneous complex 4 are also shown. The activity in the polymerization of ethylene by the complex 4 (C-1) is much higher than that of the heterogeneous system (C-2). The slower polymerization results in a higher molecular weight of the polyethylene obtained (entries 1 and 2). For propylene, the activity of the homogeneous complex depends on the solvent used. Thus, in CH_2Cl_2, a lower activity of the catalyst and lower molecular weight of the

Table 6 Results for the polymerization of ethylene and propylene with the homogeneous catalyst 4/MAO (C-1) and the heterogeneous systems 4/MAO/MCM-41 (C-2) and 4/MAO/HMS (C-3) [93]

Entry	Sample	T (°C)	P (atm)	t (h)	A^c (×10^{-4})	Type of polymer	mmmm (%)	Mw	m.p.^{d} (°C)	x^e (%)	mwd^f
Polymerization of ethylene											
1*	C-1	25	7.2	1	57.0	Solid	–	162,000^g	135	100	2.20
2^a	C-2	25	7.2	1	1.3	Solid	–	702,000	131	100	2.00
3^b	C-1	25	9.2	1	1.2	Solid	85	22,000^g	142	100	2.83
4^b	C-1	25	9.2	4	2.4	Solid	90	27,000^g	146	100	2.49
5^b	C-1	50	16	4	19.0	Solid	96	271,000^g	152	100	1.96
6^a	C-1	25	9.2	4	11.0	Solid	86	440,000	142	30	1.42
7^a	C-1	25	9.2	1	3^i	Oil	11	86,000	–	70	2.35
8^a	C-2	25	9.2	1	5.1	Elast^h	64	450,000	–	100	2.11
9^a	C-2	25	9.2	2	2.8	Elast	21.5	330,000	–	100	2.32

Table 6 (continued)

Entry	Sample	T (°C)	P (atm)	t (h)	A^c (×10^{-4})	Type of polymer	mmmm (%)	Mw	m.p.^{d} (°C)	x^e (%)	mwd^f
Polymerization of propylene
10^a	C-2	25	9.2	6	3.4	Elast	27.5	149,000	–	100	3.34
11^a	C-2	50	16	1	1.1	Elast	23.4	110,000	–	100	2.48
12^b	C-2	25	9.2	1	3.9	Oil	7.0	1,700	–	100	1.42
13^a	C-3	25	9.2	1	5.5	Solid	58	1,380,000	132	34	1.17
								290,000		56	1.53
								6,400		10	1.15
14^a	C-3	50	16	1	6.1	Elast	29.3	480,000	–	72	1.98
								3,700	–	3	1.07
								700	–	25	1.02

* The dichloride complex was used as the reference for the ethylene, see reference [18].
^{a} Toluene.
^{b} Dichloromethane.
^{c} Activity, g polymer·mol Zr^{-1}·h^{-1}.
^{d} Melting point.
^{e} Polymer fraction.
^{f} Molecular weight distribution.
^{g} Viscosity average molecular weight.
^{h} Elastomer.
^{i} The number in parentheses corresponds to the activity after one hour.

polymer are observed as compared to those in toluene (entries 3 and 6). The polymer obtained in toluene was fractionalized with the formation of a high molecular weight isotactic fraction (30%, mmmm=86%, Mw=440,000) and a low molecular weight oily fraction (70%, mmmm=11%, Mw=86,000) (entry 7). Both supported catalysts 4/MAO/MCM-41 (C-2) and 4/MAO/HMS (C-3) were found to be more active than the homogeneous complex (compare entries 6, 8, and 13). The activities of the heterogeneous catalysts, as suggested, were dependent on the type of support. The complex supported on HMS (C-3) is slightly more active than the complex on MCM-41 (C-2) (entries 8 and 13). The difference in activity increases at higher temperatures (entries 11 and 14).

There is a considerable distinction in the properties of polypropylene obtained with the homogeneous or the supported catalysts. In toluene at 25 °C, the system C-1 forms a mixture of isotactic polymer with a high molecular weight and narrow polydispersity (molecular weight distribution, mwd=1.42) and an oily atactic polymer (entry 6). The HMS-supported C-3 produces a solid polymer with lower crystallinity (mmmm=58%), while the MCM-41-supported C-2 yields an elastomeric polypropylene (entries 13 and 8). The polymer obtained with C-3 shows a multimodal molecular weight behavior: the first fraction (34%) with very high molecular weight (Mw=1,380,000) and very narrow polydispersity (1.17), the second fraction (56%) with lower molecular weight (Mw= 288,000) and slightly wider polydispersity (1.53), and a small amount (10%) of a low molecular weight third fraction (entry 13). The C-2 catalytic system produces one fraction of an elastomeric polymer (entry 8).

It is worth noting the effect of the solvent for the heterogeneous systems. In CH_2Cl_2, C-2 produces an extremely low molecular weight oily polymer as compared to those for the reaction in toluene (entries 12 and 8). It may be suggested that the presence of dichloromethane either prevents the stereoregular insertion of the monomer at the cationic active site, or induces a rapid epimerization of the last inserted unit accounting for the oily polymer. The correlation rate of these reactions influences the number and length of the isotactic domains between the stereoerrors in the polymer chain [52–59].

The differences in the properties of polypropylene obtained with homogeneous and supported catalysts are also illustrated by their ^{13}C-NMR spectra (Fig. 8). The spectrum of the polymer formed by C-1 is identical to that of isotactic polypropylene (Fig. 8a), whereas the spectra of the polymers produced by supported catalysts (Fig. 8b,c) are typical of the polypropylene in which isotactic domains alternate with many stereodefects [41].

Interesting is the large difference in activity of the supported catalysts at 60 °C and the molecular weight of the polymers obtained. The tubular 4/MAO/ MCM-41 catalyst exhibits a lower activity, producing lower molecular weight polymer as compared to those obtained at 25 °C (entries 8 and 11). This reduction in molecular weight is presumably a consequence of the rapid cleavage of the chains at higher temperatures. For the sponge-type 4/MAO/HMS catalyst, the activity was slightly increased and the molecular weight and stereoregularity of the polymer reduced (entries 13 and 14). This result again corroborates

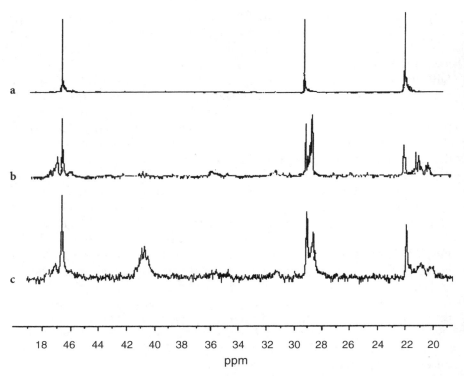

Fig. 8a–c ^{13}C-NMR spectra of polypropylene obtained with the homogeneous **C-1** (**a**) and the heterogeneous systems **C-2** (**b**) and **C-3** (**c**)

with the fact that the larger the temperature, the faster the chain termination and epimerization of the last inserted molecule of monomer.

The difference in stereoregularity of the homogeneous and supported catalysts can be explained in accordance with the presumable scheme of immobilization of the Zr benzamidinate complex on the surface of the mesoporous materials pretreated with MAO (Eq. 5). By assuming that the supported benzamidinate complex preserves its spatial conformation forming similar active species with MAO as in the solution, the formation of polypropylene with different stereospecificities may be rationalized as follows. When the benzamidinate is activated with MAO in solution, a mobile catalytic complex with an octahedral C_2 symmetry configuration (as regards the ligations) is formed. This complex provides a high rate for the propylene insertion. On the surface of the porous support, the cationic benzamidinate complex and the MAO counterion are less mobile than in solution. The rate of monomer insertion at the beginning of the reaction is the highest at the expense of a small diffusion effect. The longer the polymerization time, the less the accessibility toward the reaction center inside the pores. When the diffusion becomes the crucial factor, it causes decreases in the polymerization rate and the molecular weight of the polymers and an increase in the epimerization rate, resulting in the formation of elastomeric polypropylene with a large number of stereodefects. For the HMS support having a wormlike structure, a lower diffusion effect is expected. The data in Table 6 allow one to estimate the diffusion effect for the propylene polymerization with C-2. As may be seen, the longer the reaction time, the lower the polymerization rate, the molecular weights, and stereoregularity toward a plateau (entries 8–10). These results corroborate the envisioned diffusion effects of the MCM-41 support and the more efficient transport of the monomer molecules to the framework reaction centers for the HMS-supported [Zr benzamidinate complex.

The larger epimerization rate for the supported catalysts as compared to the homogeneous system may also be connected with the decrease of the effective positive charge of the metal center induced by the surface oxygen atoms. In favor of the last assumption are the results for the propylene polymerization with a MAO-activated isolobal Zr benzamidinate complex having a CH_2OH group at the *p*-position of the aromatic ring, instead of the CH_3 group. The presence of the OH group resulted in formation of an oily atactic polypropylene [102].

Unlike the homogeneous complex, for the supported C-3 several polymer fractions with different molecular weight, including oligomers, were detected (entries 13 and 14). This is presumably a consequence of the porous character of the supported catalysts.

3
Zirconium Allyl Complexes

Early transition metal allyl complexes have an enormous practical importance as either catalytic precursors or stoichiometric reagents in organic synthesis [103–108]. In the majority of the Group 4 complexes containing the allyl moiety, the metals exhibit the higher oxidation state (+4). Very few of these compounds are available in the literature with a +3 oxidation state, presumably because of their paramagnetic nature (reactivity) and difficulty in their handling.

Although some Ti(III)-allyl complexes have been fully characterized spectroscopically [109, 110], well-characterized Zr(III)-allyl compounds in the solid state have not been reported in the literature. On the basis of previous results obtained in our laboratory, it was very attractive and conceptually important to find a route to synthesize simple monomeric Group 4 early transition metal allyl complexes and to compare their catalytic activity to that of the well-characterized heteroallylic octahedral early transition metal compounds. Here we report the synthesis and solid-state X-ray structural characteristics of a Zr(III) bulky bis-allylic complex, and its catalytic activity in the polymerization of α-olefins [111].

The reaction of $ZrCl_4$ with two equivalents of the lithium allyl compound ($^tBuMe_2SiCH)_2CHLi \cdot TMEDA$ in toluene yielded the reduced complex [($^tBuMe_2$

SiCH)$_2$CH]$_2$Zr(III)(μ-Cl)$_2$Li·TMEDA (**8**), which was isolated as air-sensitive dark brown crystals (48%), with the concomitant formation of the dimer of the allyl ligand (**9**) in 50% yield (Eq. 6).

$$6\ ^tBuMe_2SiCH\text{---}CHSiMe_2^tBu\cdot Li\cdot TMEDA \xrightarrow[-4LiCl,\ -4TMEDA]{2\ ZrCl_4}$$

$$2\ [(^tBuMe_2SiCH)_2Zr(\mu\text{-}Cl)_2Li\cdot TMEDA]\ (\mathbf{8}) + [(^tBuMe_2Si)CH(CH=CHSiMe_2^tBu)]_2\ (\mathbf{9}) \quad (6)$$

$$(\mathbf{8}) \xrightarrow[\substack{-LiCl \\ -TMEDA \\ -SbCl_5 \\ -N(C_6H_4\text{-}Br\text{-}4)_3}]{(4\text{-}BrC_6H_4)_3NSbCl_6} (\mathbf{10})$$

The paramagnetic character of complex **8** was confirmed by ESR. The effective magnetic moment at the Zr complex (μ_{eff}=1.5±0.8 μB) was measured in toluene, indicating an unpaired electron with almost no metal–metal interaction in solution.

Complex **8** is heterodinuclear, with a Zr(III) and a Li atom bridged by two chlorine atoms (Fig. 9). The Zr(III) resides in a pseudo-octahedral environment formed by four terminal allylic carbons and two bridging chlorine atoms. The most important feature of **8** is the unsymmetrical disposition of the allylic carbons to the metal center. One of the terminal carbons in each allyl group is closer and equidistant to the Zr than the other two carbons (Zr–C9=2.361 Å, Zr–C8=2.415 Å, Zr–C7=2.474 Å, and Zr–C22=2.369 Å, Zr–C23=2.369 Å, Zr–C24=2.436 Å), creating an unbalanced C=C bond length within each allylic moiety (C9–C8=1.444 Å, C8–C7=1.393 Å, and C22–C23=1.442 Å, C23–C24= 1.385 Å). The same observation for an unsymmetrical σ,π-type (η^3-allyl) bonding has been reported for the allylic ligation in CpZr(allyl)$_3$, showing a difference of 0.182 Å between the distances of the two terminal allylic carbons from the

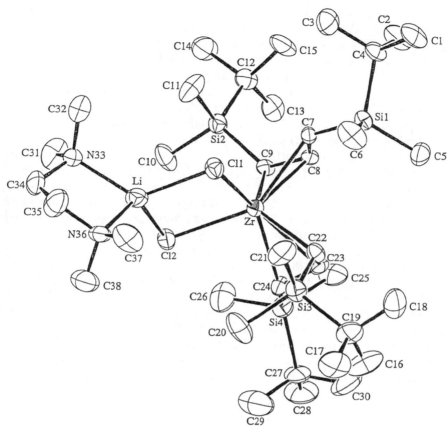

Fig. 9 ORTEP diagrams for complex **8** showing all non-hydrogen atoms [111]

metal center. In this complex the longer bond length between Zr and the terminal allylic carbon is 2.624 Å, as compared to 2.474 Å in **8**. The difference of 0.15 Å between these similar bonds in both compounds is in agreement with the expected values for a difference in the oxidation state [112].

When complex **8** is reacted in toluene in the presence of dry oxygen, an oxidation–decomposition reaction starts immediately, shown by a change in color of the reaction mixture from dark red to yellow [113, 114]. The reaction of complex **8** with equimolar amounts of $(4\text{-Br-}C_6H_4)_3NSbCl_6$ dissolved in CH_2Cl_2 yielded 23% of the complex $[(^tBuMe_2SiCH)_2CH]_2Zr(IV)Cl_2$ (**10**) as an air-sensitive light yellow powder (Eq. 6).

In the ^{1}H-NMR spectrum of **10** four different signals for each $SiMe_2$ group and each of the ^{t}Bu groups are observed. Three signals for both allylic hydrogen atoms (5.55, 5.60, 6.25 ppm) indicate that within each allylic moiety similar groups are magnetically different, although the same allylic hydrogen atoms in both ligands are in comparable magnetic environments. A corroboration of the different magnetic environments of the similar groups within each ligand in **10** is given in the ^{13}C-NMR spectra, exhibiting one signal for each SiMe$_2$ group and each of the ^{t}Bu groups. Likewise, the similarity of the magnetic environments of the allylic carbons in both ligands is confirmed by the appearance of only three signals (124.3, 125.5, and 151.8 ppm).

Although complexes **8** and **10** are in different oxidation states it seems, on the basis of the nonsymmetrical NMR results of **10** and the X-ray data of **8**, that the ligations on the Zr complexes are in similar distortions toward an unsymmetrical σ,π-type (η^3-allyl) metal complex structure [115, 116].

The reduction of the starting ZrCl$_4$ to the Zr(III) oxidation state can be explained by two parallel pathways (Scheme 4). In the first route (a), 1 equiv of the allyllithium complex (RLi) reduces the ZrCl$_4$ to ZrCl$_3$ with the concomitant formation of the allyl dimer and LiCl. The reaction of ZrCl$_3$ with 2 equiv more of the Li allyl yields complex **8**. In the second route (b), 2 equiv of the Li allyl reacts with the ZrCl$_4$ producing the bis(allyl) metal dichloride. An additional 1 equiv of Li allyl may reduce the metal, yielding complex **8** and the corresponding dimer **9**. Both mechanisms are in agreement with the yield of the complex (<50%) and with the concurrent formation of the allyl dimer. No supporting evidence for mono(allyl) or tris(allyl) complexes were detected, arguing that disproportionation reactions are not major pathways. To distinguish between both routes, the equimolar reaction of complex **10** with 1 equiv of the Li (allyl) salt was performed. The tris(allyl) complexes (observed in situ) were formed, arguing that under these reaction conditions, route (b), if operative, is not the major pathway for the production of complex **8**.

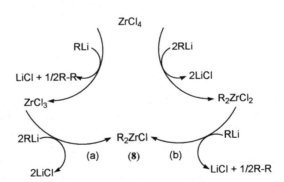

Scheme 4 Synthetic routes for the complexes **8** and **9**

Table 7 presents the results for the ethylene and propylene polymerization with the zirconium allyl complexes **8** and **10**. For both complexes, below an Al:Zr ratio of 400 neither ethylene nor propylene was polymerized. At atmospheric pressure, no polymerization of propylene was detected.

A conceptual question arises regarding the oxidation state of zirconium during the polymerization reaction. If Zr(III) is the active species, instead of Zr(IV), there should be no need for MAO as a cocatalyst. Methyl zirconium

Table 7 Results of the polymerization of α-olefins using complexes 8 and 10 as the catalytic precursors activated with MAO[a] [111]

Entry	Complex	Al:Zr ratio	T (°C)	P (atm)	Activity[b] (×10⁻⁴)	mmmm (%)	Mw	mwd[c]	m.p. (°C)
Polymerization of ethylene									
1	8	400	25	1	1.5	–	3,100,000	18.2	136.0
2	8	400	25	30	52.0	–	2,710,000	14.9	135.5
3	10	400	25	30	122.0	–	1,850,000	10.2	135.4
Polymerization of propylene									
4	8	400	25	7.2	0.7	61	42,000	2.6	136.0
5	8	800	25	7.2	1.6	56	83,000	5.3	133.0
6	8	800	50	10.1	2.1	96	29,000	2.4	152.0
7	8	1,000	25	7.2	1.3	74	46,000	2.9	138.0
8	10	1,000	25	7.2	1.2	71	44,000	2.9	138.0

[a] 2 mg catalyst (for ethylene), 5 mg catalyst (for propylene), 7 ml toluene, 1 h.
[b] g polymer·mol Zr⁻¹·h⁻¹.
[c] Molecular weight distribution.

allyl complex [(ᵗBuMe₂SiCH)₂CH]₂ZrMe obtained by alkylation of complex **8** with 1 equiv of MeLi showed no activity in the polymerization of propylene. It can be assumed that the complex of Zr(III) may undergo an oxidative coupling with the olefin followed by the generation of the active cationic moieties by MAO. Such an oxidative pathway is corroborated by the data in Table 7. As may be seen, under the same reaction conditions complexes **8** and **10** produce similar amounts of polypropylene with almost identical stereoregularities and molecular weights (compare entries 7 and 8). These results argue that only the cationic Zr(IV) allyl complexes are the active species in the α-olefin polymerizations.

For complex **8**, increasing the Al:Zr ratio results in an increase of the reaction rate and molecular weight of the polymer (entries 4 and 5). With further augmentation of the MAO concentration, a decrease in activity and molecular weight was observed. Evidently, the higher the MAO concentration, the more rapid the aluminum transfer termination pathway. This mechanism of the termination process was corroborated by the ¹³C-NMR analysis which displayed only an isobutyl signal, while vinylic carbons were not encountered. Elevating the temperature (entries 5 and 6) generates highly isotactic polymer with an increase in activity and a decrease in the molecular weight. The latter may indicate that at 50 °C, the polymer chain is cleaved from the metal center more rapidly than at 25 °C. The low molecular weight distribution for the polymer produced by the catalytic system **8**/MAO allows one to suggest that uniformity of the active sites does not change considerably during the polymerization process.

The rate of propylene polymerization with complex 8 in CH_2Cl_2 is similar to those for the reaction carried out in toluene, but an elastomeric polymer with 34% *mmmm* was obtained.

The differential scanning calorimeter studies of the polymer show that the beginning of melting signal appeared around 125–130 °C and continued up to a peak around 160 °C. The cooling curves gave reproducible results and revealed only one sharp crystallization signal within the range of 99–115.6 °C. These results indicate that within the same polymer different domains differing in their melting points are present, as supported by the existence of only one crystallization signal. The ^{13}C-NMR spectra of the polymers exhibit a high isotactic signal in addition to the other expected signals for an atactic type of polymer. The polymer shows a narrow polydispersity and the presence of only one signal on the GPC chromatograms, implying that the polymers are produced by a dynamic catalytic complex, yielding a copolymer with atactic and isotactic domains. Thus, it seems plausible that only the cationic complex 10 is able to undergo a dynamic $\eta^3 \rightarrow \eta^1$ coordination. When an η^3 coordination is operative, a cationic racemic octahedral complex, responsible for the formation of the isotactic domains, is formed. In the case of η^1 coordination, a cationic tetrahedral complex with a C_{2v} symmetry, responsible for the formation of the atactic domains, is obtained. The $\eta^3 \rightarrow \eta^1$ dynamic coordination of 10 has been experimentally corroborated [111].

For the polymerization of ethylene with complex 8, increasing the olefin pressure resulted in a considerable increase in the polymerization rate, although no major effect was observed on the molecular weight or polydispersity of the polyethylene obtained (entries 1 and 2). The high polydispersity suggests the loss of active site uniformity during the reaction. Under the same reaction conditions, complex 10 was found to be twice as active as 8, producing polyethylene with lower molecular weight and polydispersity.

4
Amido Zirconium Complexes

Replacement of the chlorine or the cyclopentadienyl ancillary ligations in metallocenes by alternative σ-donor ligands provides a possibility for changing electron-donating anionic equivalents and tailoring the steric hindrance at the metal center. We have prepared a series of soluble mono- and spirocyclic amido zirconium complexes by the incorporation of the amido ligands (R_2N-) into zirconocene, thus influencing the steric congestion and Lewis acidity at the cationic metal center [80]. The amido complexes prepared, being a new class of homogeneous polymerization catalysts, differ from the known metallocene complexes in the electron density around the metal center, thus influencing the catalytic activity in the polymerization of α-olefins.

The bis(amido) monocyclic zirconium metallocene (11) was obtained by the reaction of equimolar amounts of the zirconocene dichloride Cp_2ZrCl_2 and the dilithiated diamine $(Me_3SiN-(CH_2)_2-NSiMe_3)Li_2$ in THF (Scheme 5). (The latter

Octahedral Zirconium Complexes as Polymerization Catalysts 93

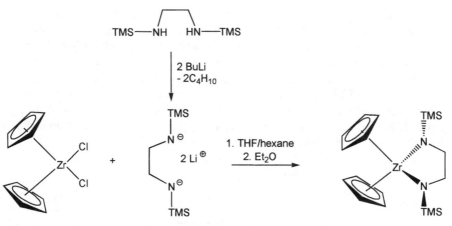

Scheme 5 Synthetic route toward the bis(amido) zirconium complex 11 [80] (**11**)

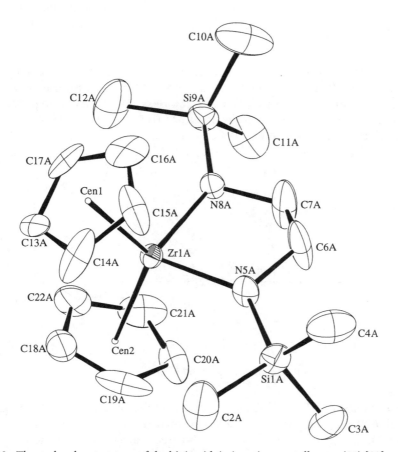

Fig. 10 The molecular structure of the bis(amido) zirconium metallocene (**11**) [80]

was prepared in situ by the reaction of N,N'-(trimethylsilyl)ethylenediamine with n-BuLi in hexane). After evaporation of the solvent and extraction into Et$_2$O and cooling (–50 °C), orange crystals of complex **11** were yielded (75%). The X-ray structure of **11** is presented in Fig. 10.

The spirocyclic tetraamido Zr complex **12** was prepared at room temperature in 82% yield by the transamination reaction of Me$_3$SiNH–(CH$_2$)$_2$–HN-SiMe$_3$ with Zr(NMe$_2$)$_4$. The latter, in turn, was obtained by the metathesis of ZrCl$_4$·2THF with four equivalents of LiNMe$_2$ (Eq. 7). Recrystallization of **12** in hexane at –75 °C yielded the pure crystalline product.

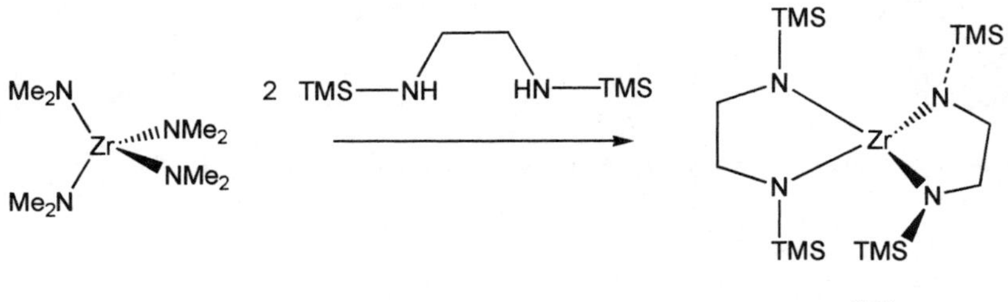

(7)

(12)

The metathesis reaction of one equivalent of the dilithiated ethylenediamine with ZrCl$_4$·2THF forms the monocyclic complex **13** (Scheme 6). Fractional crystallization from cold hexane affords a colorless microcrystalline product in 42% yield. Treatment of two equivalents of (Me$_3$SiN–(CH$_2$)$_2$–NSiMe$_3$)Li$_2$ with ZrCl$_4$·2THF results in formation of complex **12**.

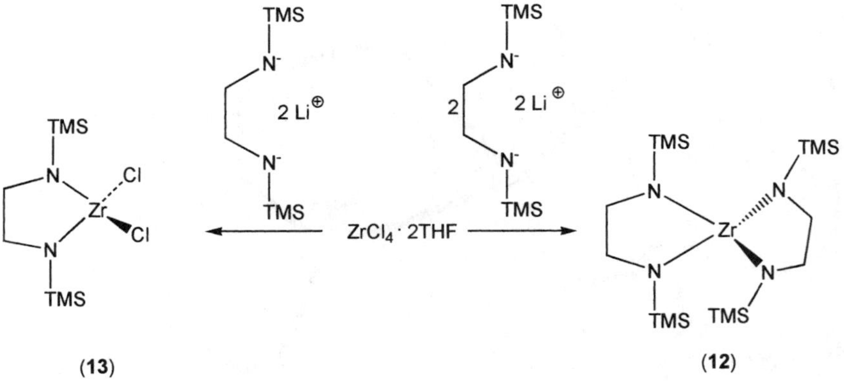

(13)   (12)

Scheme 6 Synthetic route to complexes **12** and **13** [80]

The results of ^{1}H- and ^{13}C-NMR analysis of the complexes **11–13** are given in Table 8. For comparison, the NMR resonances of the corresponding fragments of the zirconocene dichloride and free diamine used in the synthesis of amido zirconium complexes are also shown.

Octahedral Zirconium Complexes as Polymerization Catalysts 95

Table 8 The data of NMR analysis for amidozirconium complexes 11–13[a] [80]

Compound	δ^1H-NMR (ppm)			δ^{13}C-NMR (ppm)		
	Cp group	Ethylene bridge	Me$_3$Si group	Cp group	Ethylene bridge	Me$_3$Si group
11	6.14	3.29	0.07	112.9	54.5	1.0
12		3.74	0.07		54.1	0.3
13		3.61			69.3	
Cp$_2$ZrCl$_2$	6.47			116.0		
Free diamine		2.48	–0.14		45.4	–0.2

[a] NMR analysis was performed in benzene-d_6 at 25 °C.

As can be seen, shifts of the Cp group signals to higher field values are observed in the ^{1}H- and ^{13}C-NMR spectra in passing from complex **11** to the zirconocene dichloride. This may be a consequence of the increased electron density at the cyclopentadienyl ancillary ligand due to the electron-donor capacity of the chelating bis(amido) ligand compared with the two chlorine atoms in Cp$_2$ZrCl$_2$. Thus, one can expect that the Zr–N bond is less strong than the Zr–Cl bond, allowing activation of the metallocene amido complex **11** by Lewis acids. Comparison of the ^{1}H- and ^{13}C-NMR resonances related to the ethylene bridges for the bis(amido) ligand in **11** and in the free diamine showed the lower field shifts in **11** caused by the attachment of the bis(amido) ligand to the Lewis acid Zr atom. The ^{1}H- and ^{13}C-NMR signals of the methyl resonances of the SiMe$_3$ groups in **11** appear to have a similar downfield trend, compared with the corresponding signals of the free diamine.

The electron transfer from the two bis(amido) ligands toward the electrophilic Zr atom in complexes **12** and **13** is ascertained by the chemical shifts of the ethylene groups. A similar trend in **12** is observed for the Me$_3$Si groups. These results, compared with those for complex **11**, show a similar electronic environment between the Cp and the bis(amido) ancillary ligands.

Comparison of the thermodynamic bond disruption energies in Cp$_2$ZrCl$_2$ and complex **11** ($D_{(Zr-Cl)}$=491 kJ mol^{-1}, $D_{(Zr-N)}$=355 kJ mol^{-1}), and our previous observation that early/late phosphido- and arsenido-bridged heterobimetallic complexes can be activated by strong Lewis acids to produce the methyl cationic complex [79, 117], suggested that, in a similar fashion, it would be possible to activate heterolytically complexes **11–13** by MAO forming an active species for the polymerization of α-olefins (Eq. 8). The cationic complex is, presumably, formed by the double metathesis of the amido ligands and the methyl groups of MAO in a similar way as found for metallocene phosphido ligands [80].

The zirconocene complex **11**, when activated with methylalumoxane, shows high activity in ethylene polymerization (Table 9) which is slightly larger than

Table 9 Activity of the amido zirconium complexes 11–13 for the polymerization of olefins[a] [80]

Entry	Catalyst	[Catalyst] (mmol)	MAO/catalyst Ratio	Activity[b] (×10⁻³)
Polymerization of ethylene				
1	11	3.54	1,000:1	2,280
2	11	3.54	3,000:1	7,530
3	12	3.02	1,000:1	3.61
4	12	3.02	3,000:1	1.08
5	13	5.49	1,000:1	7.23
6	13	5.49	3,000:1	5.68
Polymerization of propylene				
7	11	3.54	1,000:1	1,090
8	11	3.54	3,000:1	1,220

[a] In 50 ml of toluene, 25 °C, 1.0 atm.
[b] g polymer·mol Zr⁻¹·h⁻¹.

(8)

the activity of the catalytic system Cp$_2$ZrCl$_2$/MAO [7, 118]. The activity of **11** increases with increasing Al:Zr ratio (entries 1 and 2). Complexes **12** and **13** were found to be less active than the corresponding complex **11**. Furthermore, in contrast to complex **11**, the catalysts **12** and **13** show the opposite trend of the relationship between activity and MAO/catalyst ratio. Such behavior may be a result of the large coordinative unsaturation and electrophilicity of the metal center, which promotes the active "cationic" complexes to coordinate with the solvent or MAO [7].

The catalytic system **11**/MAO shows lower activity in propylene polymerization as compared to polymerization of ethylene, with a similar trend in the relationship between the polypropylene yield and Al:Zr ratio (compare entries 1,2 7, and 8). The proton and carbon spectroscopic analysis of the polypropylene obtained revealed only vinyl/isopropyl end groups. The fact that no vinylidene/*n*-propyl end groups were observed testifies that polypropylene is formed in accordance with a β-methyl elimination pathway.

Thus, we have shown that, in addition to the known methyl and chloride ancillary ligands in zirconocenes, amido ligands can also be activated by strong Lewis acids such as MAO, producing cationic complexes active in the polymerization of α-olefins.

5
Zirconium-Containing Homoleptic Phosphinoamide Dynamic Complexes

Investigations of the bis(benzamidinate) dichloride or dialkyl complexes of Group 4 metals show that these complexes, obtained as a racemic mixture of *cis*-octahedral compounds with C_2 symmetry, are active catalysts for the polymerization of α-olefins when activated with MAO or perfluoroborane cocatalysts [29–41]. As was demonstrated above, polymerization of propylene with these complexes at atmospheric pressure results in the formation of an oily atactic product, instead of the expected isotactic polymer. The isotactic polypropylene (*mmmm*>95%, m.p.=153 °C) is formed when the polymerization is carried out at high concentration of olefin (in liquid propylene), which allows faster insertion of the monomer and almost completely suppresses the epimerization reaction.

A conceptual question is whether simple octahedral or even tetrahedral complexes that have a dynamic Lewis-basic pendant group, donating a pair of electrons to the metal center, are suitable for the production of an elastomeric polypropylene. As shown in Scheme 7, a dynamic equilibrium may take place between a tetrahedral and an octahedral configuration (X=halide, E=donor group with a lone electron pair, R=C, N, P, or other anionic bridging group). (A plausible *trans*-octahedral complex, which can be formed in this type of dynamic process, is unable to perform the olefin insertion and has no catalytic activity [5, 20, 80, 81].)

To check the presented hypothesis, the Zr-containing homoleptic phosphinoamide [Zr(NPhPPh$_2$)$_4$] (**14**) was prepared by reacting ZrCl$_4$ with four equivalents of LiNPhPPh$_2$ (Eq. 9) [119]. The latter was synthesized by the deprotonation of phosphinoamine Ph$_2$PN(H)Ph with butyllithium in hexane [120]. At room temperature, two signals are observed in the ^{31}P-NMR spectrum of **14** (24.7, s, and

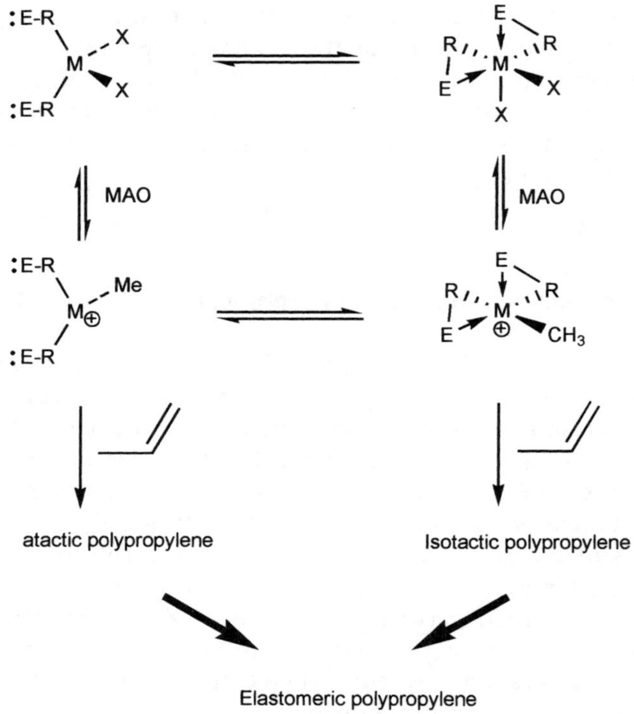

Scheme 7 Elastomeric polypropylene obtained by dynamic interconversion between cationic complexes with C_{2v} and C_2 symmetry

–5.4 ppm, br, in THF), which are shifted to high field by about 4–43 ppm relative to the starting Ph$_2$PN(H)Ph (28.6 ppm). At lower temperature, the signal at –5.4 ppm sharpens and the ratio of the two signals changes from 1:1 at 253 K to 1:3 at 223 K. Apparently, at 253 K two of the four phosphino groups exhibit η^1 coordination in solution (Eq. 9).

The polymerization of propylene using complex **14** activated by MAO (Al:Zr ratio=500, solvent toluene, 25 °C) yielded 80 g polymer·mol Zr^{-1}·h^{-1} with a molecular weight Mw=115,000 and polydispersity=2.4 [119]. The reaction was carried out in liquid propylene to avoid, as much as possible, the epimerization of the last inserted monomer unit and to allow rational design of the elastomeric polymer. The formation of elastomeric polypropylene is consistent with the proposed equilibrium between *cis*-octahedral cationic complexes with C_2 symmetry inducing the formation of the isotactic domain, and tetrahedral complexes with C_{2v} symmetry responsible for the formation of the atactic domain (Scheme 7). The narrow polydispersity of the polypropylene obtained supports the polymerization mechanism in which the single-site catalyst is responsible for the formation of the elastomeric polymer.

The ^{13}C-NMR spectrum of the polymer (Fig. 11) exhibits in the methyl region a large signal of the *mmmm* pentad, ruling out the possibility of for-

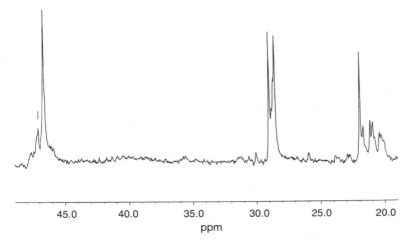

Fig. 11 ^{13}C-NMR spectrum of the elastomeric polypropylene prepared with the catalytic system 14/MAO [119]

mation of a high molecular weight atactic polypropylene and indicating polymer molecules with a combination of atactic and isotactic blocks and a large domain of isotactic pentads [121–123]. No vinylidene or vinyl chain end groups were detected in the ^{1}H-NMR spectrum, suggesting an aluminum chain transfer mechanism as the major termination step.

To corroborate that the epimerization reaction is responsible for the stereo-errors in the polypropylene chain, we treated 1-octene with complex 14 activated with MAO. The reaction resulted in the quantitative formation of *trans*-2-octene. In addition, the reaction of allylbenzene with the system 11/MAO at room temperature produced 100% conversion of the former to *trans*-methylstyrene. The isomerization results indicate that during the polymerization, the metal center in 14 is also able to migrate through the growing polymer chain, inducing branching and additional environments for the methyl ligands (Scheme 2) [124].

Thus, simple tetrahedral complexes with ancillary ligands incorporating donating groups (with a lone pair of electrons able to coordinate to the metal center) allow the selective design of elastomeric polypropylenes.

6
The Structure of the Elastomeric Polypropylene

As already mentioned, on the polymerization of propylene at higher monomer concentrations using zirconium-containing octahedral complexes with a C_2 symmetry, an elastomeric polypropylene was formed. It was essential to compare the chemical structure of the obtained elastomers with the structure of the elastomeric polypropylenes described in the literature.

Three types of elastomeric polypropylene are known: (1) a large molecular weight atactic polymer [125], (2) polymers with alternating isotactic and atactic blocks [108, 109, 126–129], and (3) polymers formed by the "dual-site" mechanism [110]. The structure of the obtained elastomeric polypropylene was elucidated by comparison of the ^{13}C-NMR spectra with those of an atactic oily polymer and isotactic polypropylene prepared by the zirconium complex 4. On the basis of the NMR data, for all the samples the statistical lengths of the isotactic blocks between two neighboring epimerization stereodefects were calculated [41]. It was found that for the isotactic polymer (*mmmm*~90%), the length of an isotactic block before a stereodefect was 35–45 CH$_3$ groups, while for the atactic sample (*mmmm*~7%) it was 7 or 8 CH$_3$ groups. For the elastomeric polypropylene obtained with the zirconium benzamidinate complex (4), the length of the isotactic fragment was found to be between 11 and 27 methyl groups [41]. Thus, for the elastomeric polymer the length of the statistical isotactic fragment is shorter than that for an isotactic polymer and longer than that for an atactic sample.

It is plausible to assume that the elastomeric polypropylene obtained differs, in principle, from the above-mentioned types of elastomers. Contrary to the high molecular weight atactic elastomers [125], our samples have an isotactic structure in which longer isotactic domains are disposed between stereodefects. Differing from the alternating isotactic–atactic block elastomers [108, 109, 126–129], where the alternating domains have commensurate lengths, our polymers are distinguished by frequent alternation of the isotactic fragments with many stereodefects. As compared to the elastomers of type (3), which exhibit large *mmmm* contents (up to 72%) [110], our samples are characterized by lower stereoregularity. Besides, in the dual-site mechanism each site has a different symmetry and, by the back-skipping of the polymer chain, two stereochemistries for the insertion are possible. With our complexes, back-skipping of the polymer chain induces the same symmetry, thus inducing the same type of stereoregular insertion. The stereoregular errors are formed because of two competing processes, the insertion of propylene and the intramolecular epimerization of the growing chain at the last inserted unit. The rate correlation of these processes influences the number and length of the isotactic domains between the stereoerrors.

To discriminate between the crystalline and amorphous areas in the homogeneous polymer, we have utilized atomic force microscopy (AFM). Figure 12 depicts typical phase images of isotactic, elastomeric, and atactic polypropylene samples. The sample of the isotactic polymer consists of a mosaic of crystalline islands (brighter spots in the image) embedded in an amorphous phase (darker areas) of the same material (Fig. 12a). The elastomeric polymer is composed of long interpenetrated and branched fibrils of crystalline character that interconnect amorphous regions (Fig. 12b). This phase image provides further support to the concept of microheterogeneous crystallinity of the elastomeric phase, caused by different crystalline domains that are characterized by slightly different melting points. This set of heterogeneous crystalline phases is claimed

Octahedral Zirconium Complexes as Polymerization Catalysts

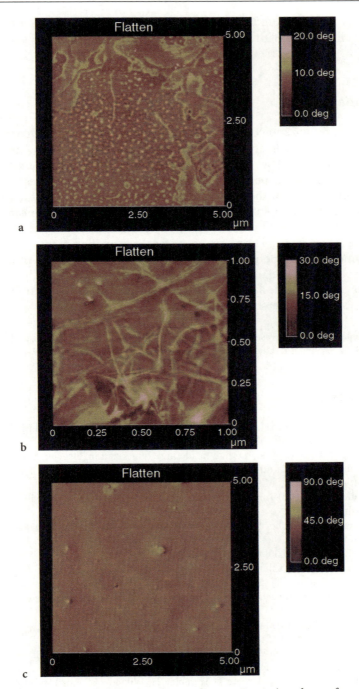

Fig. 12a–c AFM pictures of different polypropylenes. **a** Isotactic polymer from Table 2, entry 4; **b** elastomeric polymer obtained with Ti benzamidinate complex [41]; **c** atactic polymer fraction from Table 2, entry 8

to yield a broad and undefined melting zone. The atactic polymer is predominantly amorphous with a low density of crystalline domains (Fig. 12c).

7
Conclusion

The synthesis and molecular structure of several octahedral zirconium complexes with various ligations alternative to the cyclopentadienyl ligands, and their catalytic properties in the polymerization of α-olefins, are described. It is demonstrated that, depending on the structural features of these complexes and the conditions of the polymerization process (amount of the complexes and cocatalysts, nature of the solvents used, temperature, and concentration of the monomers), highly stereoregular, elastomeric, or oily atactic polypropylenes can be produced. The mechanisms for the formation of the different types of polymers are competitive reactions inhibiting the formation of the stereoregular polymer. The data presented in this review shed light on the conceptual question regarding the applicability of cis-octahedral C_2-symmetry complexes as catalysts for the stereospecific polymerization of α-olefins.

Acknowledgements The research was supported by the German–Israel Foundation grant no. I-621-27.5/1999. A.L thanks the Ministry of Immigration for the KAMEA Fellowship.

References

1. Ziegler K, Holzkamp H, Breil H, Martin H (1955) Angew Chem Int Ed Engl 67:54
2. Natta D (1956) Angew Chem Int Ed Engl 68:393
3. Keii T, Soga K (eds) (1990) Catalytic olefin polymerization. Elsevier, Amsterdam
4. Kaminsky W (2002) Adv Catal 46:89
5. Kaminsky W, Arndt M (1977) Adv Polym Sci 127:143
6. Mulhaupt R (1999) In: Karger-Kocsis J (ed) Polypropylene: A–Z reference. Kluwer, Dordrecht, p 454
7. Brintzinger H, Fischer D, Multhaupt R, Rieger B, Waymouth RM (1995) Angew Chem Int Ed Engl 34:1143
8. Alt HG, Köpll A (2000) Chem Rev 100:1205
9. Coates GW (2000) Chem Rev 100:1223
10. Chen EY-X, Marks TJ (2000) Chem Rev 100:1391
11. Metz MV, Schwartz DJ, Stem CL, Nickias PN, Marks TJ (2000) Angew Chem Int Ed 39:1312
12. Lanza G, Fragalá IL, Marks TJ (2000) J Am Chem Soc 122:12764
13. Ittel SD, Johnson LK, Brookhardt M (2000) Chem Rev 100:1169
14. Britovsek GJP, Gibson VC, Wass DF (1999) Angew Chem Int Ed 38:428
15. Bochmann M, Lancaster SJ (1993) Organometallics 12:663
16. Ewart SW, Sarsfield MJ, Jeremic D, Tremblay TL, Williams EF, Baird MC (1998) Organometallics 17:1502
17. Pellecchia C, Immirzi A, Grassi A, ZambelliA (1993) Organometallics 12:4473

18. Shrock RR, Baumann R, Reid SM, Goodman JT, Stumpf R, Davis WM (1999) Organometallics 18:3649
19. Mack H, Eisen MS (1998) J Chem Soc Dalton Trans 917
20. Gielens EECG, Tiesnitsch JY, Hessen B, Teuben JH (1998) Organometallics 17:1652
21. Chen Y-X, Fu P-F, Stern CL, Marks TJ (1997) Organometallics 16:5958
22. Kempe R (2000) Angew Chem Int Ed 39:468
23. Goodman JT, Schrock RR (2001) Organometallics 20:5205
24. Schrock RR, Bonitatebus PJ, Schrodi Y (2001) Organometallics 20:1056
25. O'Connor PE, Morrison DJ, Steeves S, Burrage K, Berg DJ (2001) Organometallics 20:1153
26. Yoshida Y, Matsui S, Takagi Y, Mitani M, Nakano T, Tanaka H, Kasiwa N, Fujita T (2001) Organometallics 20:4793
27. Layrante KC, Sita LR (2001) J Am Chem Soc 123:10754
28. Gomez R, Duchateau R, Chernega AN, Teuben JN, Edelmann FT, Green MLH (1995) J Organomet Chem 491:153
29. Hagadorn JR, Arnold J (1998) Organometallics 17:1355
30. Edelmann FT (1996) Top Curr Chem 179:113
31. Herscovics-Korine D, Eisen MS (1995) J Organomet Chem 503:307
32. Richter J, Edelmann FT, Noltemeyer M, Schmidt H-G, Shmulinson M, Eisen MS (1998) J Mol Catal 130:149
33. Volkis V, Shmulinson M, Averbuj C, Lisovskii A, Edelmann TF, Eisen MS (1998) Organometallics 17:3155
34. Walter D, Fischer R, Friedrich F, Gebhardt P, Görls H (1996) J Organomet Chem 508:13
35. Walter D, Fischer R, Görls H, Koch J, Scheweder B (1996) J Organomet Chem 508:13
36. Duchateau R, van Wee CT, Meetsma A, van Duijnen PT, Teuben JN (1996) Organometallics 15:2279
37. Gomez R, Green MLH, Haggit JL (1996) J Chem Soc Dalton Trans 939
38. Flores JC, Chien JCW, Rausch MD (1995) Organometallics 14:1827
39. Flores JC, Chien JCW, Rausch MD (1995) Organometallics 14:2106
40. Gomez R, Ducheteau R, Chernega AN, Meetsma A, Edelmann FT, Teuben JH, Green MLH (1995) J Chem Soc Dalton Trans 217
41. Volkis V, Nekkenbaum E, Lisovskii A, Hasson G, Semiat R, Kapon M, Botoshansky M, Eishen Y, Eisen MS (2003) J Am Chem Soc 125:2179
42. Kapon M, Eisen MS (1994) J Chem Soc Dalton Trans 3507
43. Fenske D, Hartmann E, Dehnicke K (1988) Z Naturforsch 43b:1611
44. Hagadorn JR, Arnold J (1997) J Chem Soc Dalton Trans 3087
45. Sinn H, Kaminsky W (1980) Adv Organomet Chem 18:99
46. Bochmann M, Cuenca T, Hardy DT (1994) J Organomet Chem 484:C10
47. March J (1985) In: Advanced organic chemistry, 3rd edn. Wiley-Interscience, New York
48. Resconi L, Piemontesi F, Franciscono G, Abis L, Fiorani T (1992) J Am Chem Soc 114:1025
49. McCord EF, McLain SJ, Nelson LTJ, Arthur SD, Coughlin EB, Ittel SD, Johnson IK, Tempel D, Killian CM, Brookhart M (2001) Macromolecules 34:362
50. Busico V, Cipillo R, Caporaso P, Angeloni G, Segre AL (1998) J Mol Catal 128:53
51. Bisico V, Cipillo R, Monaco G, Vacatello M (1997) Macromolecules 30:6251
52. Busico V, Brita D, Caporaso L, Cipillo R, Vacatello M (1997) Macromolecules 30:3971
53. Busico V, Caporaso L, Cipillo R, Landriani L, Angelini G, Margonelli A, Segre AL (1996) J Am Chem Soc 118:2105
54. Leclerc M, Brintzinger HH (1996) J Am Chem Soc 118:9024
55. Busico V, Cipillo R, Corradini P, Landriani L, Vacatello M, Segre AL (1995) Macromolecules 28:1887
56. Busico V, Cipillo R (1994) J Am Chem Soc 116:9329

57. Resconi L (1999) J Mol Catal 146:167
58. Yang X, Stern CL, Marks TJ (1994) J Am Chem Soc 116:10015
59. Giardello MA, Eisen MS, Stern CL, Marks TJ (1993) J Am Chem Soc 115:3326
60. Lancaster SJ, Robinson OB, Bochmann M, Coles SJ, Hursthouse MB (1995) Macromolecules 14:2456
61. Gillis DJ, Tudoret MJ, Baird MC (1993) J Am Chem Soc 115:2543
62. Eisen MS, Marks TJ (1992) J Am Chem Soc 114:10358
63. Eisen MS, Marks TJ (1992) Organometallics 11:3939
64. Ducheteau R, Brussee EAC, Meetsma A, Teuben JN (1997) Organometallics 16:5506
65. Duchateau R, Tuinstra T, Brussee EAC, Meestma A, van Duijnen PT, Teuben JH (1997) Organometallics 16:3511
66. Averbuj C, Eisen MS (1999) J Am Chem Soc 121:8755
67. Eshuis JJW, Tan YY, Teuben JH, Renkema J (1990) J Mol Catal 62:277
68. Resconi L, Giannini U, Albizzati E, Piemontesi F, Fiorani T (1991) Polymer Prepr 32:463
69. Jia L, Yang X, Ishihara A, Marks TJ (1995) Organometallics 14:3135
70. Jordan RF, Bradley PK, LaPointe RE, Taylor DF (1990) New J Chem 14:499
71. Shapiro PJ, Cotter WD, Schaefer WP, Labinger JA, Bercaw JE (1994) J Am Chem Soc 116:4623
72. Bierwagen EP, Bercaw JE, Goddard WA (1994) J Am Chem Soc 116:1481
73. Kuribayashi HK, Koga N, Morocuma K (1992) J Am Chem Soc 114:2359
74. Averbuj C, Tish E, Eisen MS (1998) J Am Chem Soc 120:8640
75. Edelmann FT (1994) Coord Chem Rev 137:403
76. Lambert JB (1990) Tetrahedron 46:2677
77. Apeloig Y, Biton R, Freih AA (1993) J Am Chem Soc 115:2522
78. Blumenkopf TA, Overman LE (1986) Chem Rev 86:857
79. Shribman T, Kurz S, Senff U, Lindenberg F, Hey-Hawkins E, Eisen MS (1998) J Mol Catal A Chem 129:191
80. Mack H, Eisen MS (1996) J Organomet Chem 525:81
81. Bochmann M (1996) J Chem Soc Dalton Trans 255
82. Lechler R, Hausen H.-D, Weidlein J (1989) J Organomet Chem 359:1
83. Duchateau R, Meetsma A, Teuben JH (1996) J Chem Soc Dalton Trans 223
84. Metz MV, Schwartz DJ, Stern CL, Marks YJ, Nickias PN (2002) Organometallics 21:4159
85. Metz MV, Sum Y, Stern CL, Marks TJ (2002) Organometallics 21:3691
86. Jia L, Yang X, Stern CL, Marks TJ (1997) Organometallics 16:842
87. Charoenchaidet S, Chavadej S, Gulari E (2002) J Polym Sci A 40:3240
88. Misikabhumma K, Spaniol TP, Okuda J (2002) J Macromol Chem Phys 203:115
89. Galan-Fereres M, Koch T, Hey-Hawkins E, Eisen MS (1999) J Organomet Chem 580:145
90. Arrowmith D, Kaminsky W, Schauwienold A.-M, Weingarten U (2000) J Mol Catal A Chem 160:97
91. Soga K (1993) Macromol Chem 66:43
92. Hagimoto H, Shiono T, Ikeda T (2002) Macromolecules 35:5744
93. Ray B, Volkis V, Lisovskii A, Eisen MS (2002) Israel J Chem 42:333
94. Beck JC, Vartuli JC, Roth WJ, Leonowicz ME, Kresge CT, Schmitt KD, Chu CT.-W, Olson DH, Shepard EW, McCullen SB, Higgins JB, Schlenker JL (1992) J Am Chem Soc 114:10834
95. Tanev PT, Pinnavaia TJ (1996) Chem Mater 8:2068
96. Ryoo R, Kim JM (1995) J Chem Soc Chem Commun 711
97. Bagshaw SA, Prouzet E, Pinnavaia TJ (1995) Science 269:1242
98. Zhang W, Pauly TR, Pinnavaia TJ (1997) Chem Mater 9:2491
99. Pauly TR, Liu Y, Pinnavaia TJ, Billinge SJL, Rieker TP (1999) J Am Chem Soc 21:8835

100. Chen C-Y, Li H-X, Davis ME (1993) Microporous Mater 2:17
101. Tismaneanu R, Ray B, Khalfin R, Semiat R, Eisen MS (2001) J Mol Catal A Chem 171:229
102. Chen F, Kapon M, Eisen MS (manuscript in preparation)
103. Wike G (1963) Angew Chem 75:10
104. Davies SG (1982) Organotransition metal chemistry: application to organic synthesis. Pergamon, Oxford
105. Trost BM (1986) J Organomet Chem 300:263
106. Perry DC, Farson FS, Schoenberg E (1975) J Polym Sci 13:1071
107. Bergbreiter DE, Parsons GL (1981) J Organomet Chem 208:47
108. Klei E, TeubenJH, de Liefde Meijer HJ, Kwak EJ, Bruins AP (1982) J Organomet Chem 224:327
109. Chen Y, Kai Y, Kasai N, Yashida H, Yamamoto H, Nakamura A (1981) J Organomet Chem 407:191
110. Martin HA, Jellinek F (1967) J Organomet Chem 8:115
111. Ray B, Gueta Neyroud T, Kapon M, Eishen Y, Eisen MS (2001) Organometallics 20:3044
112. Shannon RD (1976) Acta Crystallogr A32:751
113. Meijere A, Stecker B, Kourdoiukov A, Williams CM (2000) Synthesis 7:929
114. Lin X, Novak BM (2000) Macromolecules 33:6205
115. Erker G, Berg K, Angermund K, Krüger C (1987) Organometallics 6:2620
116. Erker G, Dorf U, Benn R, Reinhardt R-D, Peterson JL (1984) J Am Chem Soc 106:7649
117. Lindenberg F, Shribman T, Sieler J, Hey-Hawkins E, Eisen MS (1996)
118. Brown SJ, Gao X, Harrison DG, Koch L, Spence REGH, Yap GPA (1998) Organometallics 17:5445
119. Kühl O, Koch T, Somoza FB Jr, Junk PC, Hey-Hawkins E, Plat D, Eisen MS (2000) J Organomet Chem 604:116
120. Ashby MT, Li Z (1992) Inorg Chem 31:1321
121. Coates GM, Waymouth RM (1995) Science 267:217
122. Hu Y, Krejchi MT, Shah CD, Myers CI, Waymouth RM (1998) Macromolecules 31:6908
123. Dietrich U, Haksmann M, Rieger B, Klinga M, Leslelä M (1999) J Am Chem Soc 121:4348
124. Bovey FA, Mirau PA (eds) (1996) NMR of polymers. Academic, San Diego
125. Resconi L, Jones RL, Rheingold AL, Yap GPA (1996) Organometallics 15:998
126. Tagge CD, Kravchenko RL, Lal TK, Waymouth RM (1999) Oragnometallics 18:380
127. Maciejewski Petoff JL, Agoston T, Lal TK, Waymouth RM (1998) J Am Chem Soc 120:11316
128. Kravchenko R, Massod A, Waymouth RM, Myers CL (1998) J Am Chem Soc 120:2039
129. Bruce MD, Waymouth RM (1998) Macromolecules 31:2707

Topics Organomet Chem (2005) 10: 107–132
DOI 10.1007/b98422
© Springer-Verlag Berlin Heidelberg 2005

Zirconocene Complexes as New Reagents for the Synthesis of Cyclopropanes

Jan Szymoniak (✉) · Philippe Bertus

UMR 6519 – Réactions Sélectives et Applications, CNRS –
Université de Reims Champagne-Ardenne, BP 1039, 51687 Reims Cedex 2, France
jan.szymoniak@univ-reims.fr

1	Introduction	108
2	Construction of the Cyclopropane Ring from Two- and One-Carbon Synthons	109
3	Construction of the Cyclopropane Ring from Three-Carbon Synthons	114
3.1	Formation of Cyclopropanes via γ-Elimination Reactions	114
3.1.1	Zirconacycles as Intermediates	114
3.1.2	Reactions Initiated by Hydrozirconation	117
3.2	Cyclopropanes from Homoenolate Anion Equivalents	123
4	Miscellaneous Cyclopropane Syntheses	125
5	Conclusion	129
	References	130

Abstract The use of zirconocene chemistry for the synthesis of cyclopropanes is a recent research area. Most of the reactions in this chapter have been reported in the last few years and not entirely explored. Yet, cyclopropanes were formed from nonconventional starting materials (other than alkenes and carbenoids) in various ways. Cyclopropanes can be constructed from two- and one-carbon building blocks, using organomagnesium reagents and carbonyl compounds, through a zirconium-mediated deoxygenative coupling reaction. The γ-elimination processes from three-carbon zirconocene intermediates open selective pathways to cyclopropanes from readily available starting materials such as vinyl epoxides, allylic ethers, homoallylic ethers, and bromides. An easy transmetalation from zirconium to other metals, such as Al or Zn, expands the number of efficient methods for the construction of cyclopropanes. Based on zirconocene chemistry, synthetically useful cyclopropanation reactions might be further developed.

Keywords Cyclopropane · Lewis acid · Metallacycles · Zirconium

1
Introduction

Not only do cyclopropanes appear as versatile synthetic intermediates [1], but also the cyclopropane framework is present in a number of biologically active natural and synthetic molecules [2]. For these reasons, continuous efforts have been made over the last few years to develop new methods for cyclopropane construction [3]. Among them, those employing transition metal-mediated processes play a leading role. Selective and especially enantioselective cyclopropanations through decomposition of diaza compounds with Cu and Rh, as well as Simmons–Smith-type reactions, should be noted [4]. Another spectacular recent development has been achieved in the area of cyclopropane synthesis through titanium chemistry. Unlike conventional cyclopropanation reactions employing alkenes or equivalents, the Kulinkovich reaction (1989) provides a general pathway to the synthesis of cyclopropanols from carboxylic esters, organomagnesium reagents, and readily available Ti(IV) reagents (typically Ti(Oi-Pr)$_4$) (Scheme 1) [5].

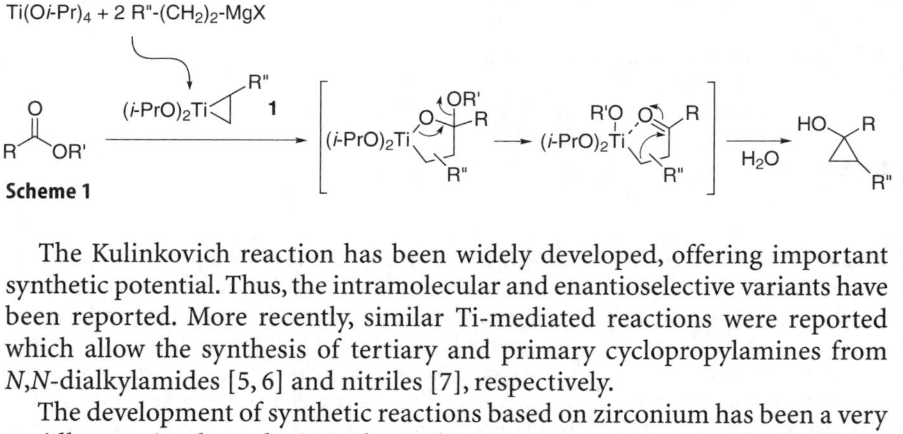

Scheme 1

The Kulinkovich reaction has been widely developed, offering important synthetic potential. Thus, the intramolecular and enantioselective variants have been reported. More recently, similar Ti-mediated reactions were reported which allow the synthesis of tertiary and primary cyclopropylamines from N,N-dialkylamides [5, 6] and nitriles [7], respectively.

The development of synthetic reactions based on zirconium has been a very rapidly growing branch since the mid-1980s. In particular, Zr(II) complexes related to the zirconocene (Cp$_2$Zr) series display various and synthetically useful transformations [8]. A comparison of Ti(II) and Zr(II) chemistry reveals similar reaction patterns but also striking differences. It offers opportunities for developing complementary approaches to controlling reactivity and selectivity. In this chapter, we focus on the cyclopropane ring-forming reactions that involve zirconocene intermediates. Most of these reactions have only recently been reported. They would make a basis for new synthetic methods leading to cyclopropanes.

The reactions presented in this chapter are divided into three groups: deoxygenative processes employing two-carbon and one-carbon synthons (Sect. 2), reactions employing three-carbon synthons (Sect. 3) and miscellaneous syntheses, involving indirect zirconium assistance (Sect. 4).

2
Construction of the Cyclopropane Ring from Two- and One-Carbon Synthons

A deoxygenative conversion of carbonyl compounds into cyclopropanes (Scheme 2) stretches the bounds of the conventional cyclopropanation reactions that employ the C=C double bond and a carbenoid intermediate.

Scheme 2

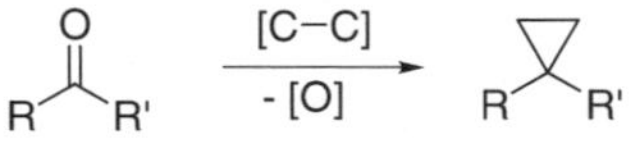

In fact, only in some cases are real carbenoid species involved in these reactions. The reactions employing oxophilic metals such as Zn, Fe, Sm, and In were reported [9], and representative examples are given in Scheme 3. Such reactions exhibit obvious synthetic limitations, since only aromatic (unsaturated) carbonyl compounds can be used, and/or cyclopropanes of a very specific structure obtained. The Kulinkovich and related reactions, involving dianion equivalents (1, Scheme 1), represent the only synthetically useful deoxygenative process at present.

In this context, zirconium chemistry could open up new prospects for developing deoxygenative approaches to cyclopropanes. The Zr-assisted variant of the Kulinkovich reaction has been reported (Scheme 4, Eq. 1) [10]. Cyclopropanols were also formed, as by-products in addition to homoallylic alcohols, from aliphatic acid chlorides (Eq. 2) [10]. The conversion of acid chlorides (and also esters in several cases) into the corresponding homoallylic alcohols

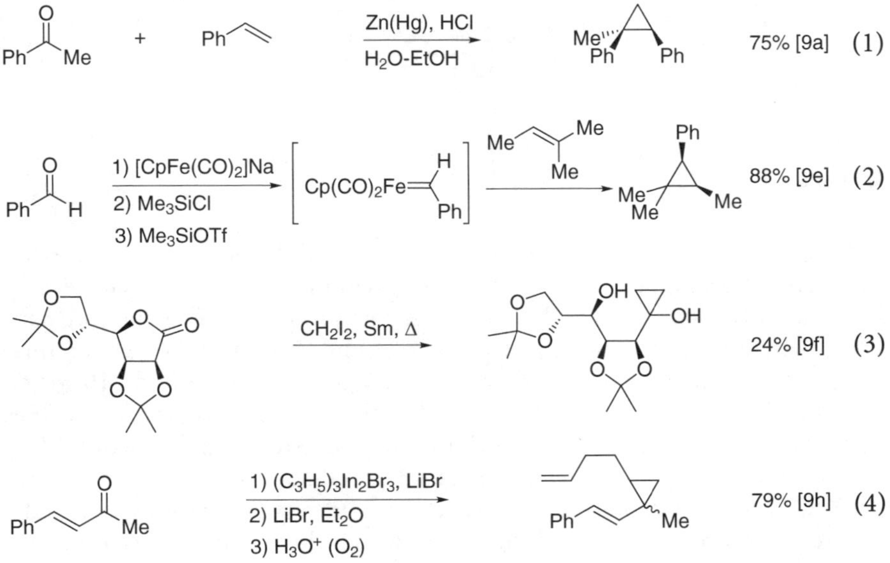

Scheme 3

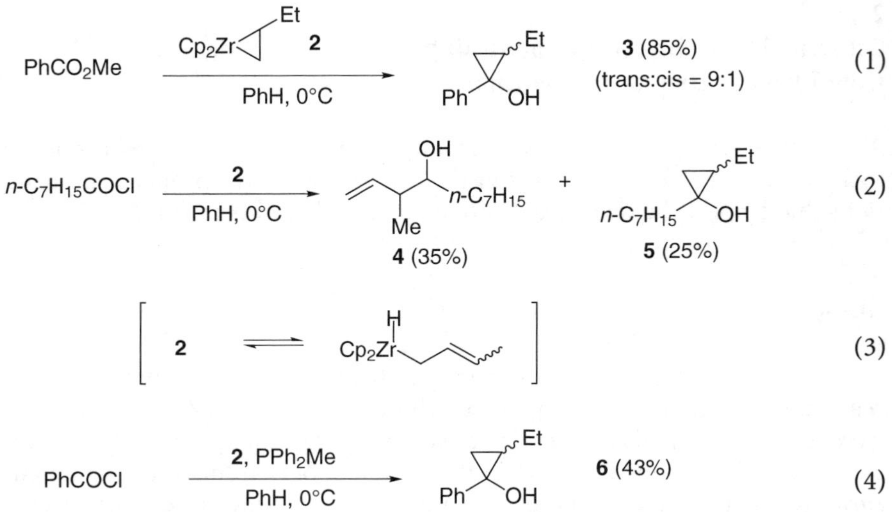

Scheme 4

(**2** is formed in situ from Cp_2ZrCl_2 and 2 n-BuLi, see: [8])

would proceed through the allylic C–H bond activation (Eq. 3). It could then be avoided in the presence of a stabilizing phosphine as additive (Eq. 4). Finally, the cyclopropanol **8** was formed from the glucone **7** in the presence of a Lewis acid (Scheme 5) [11].

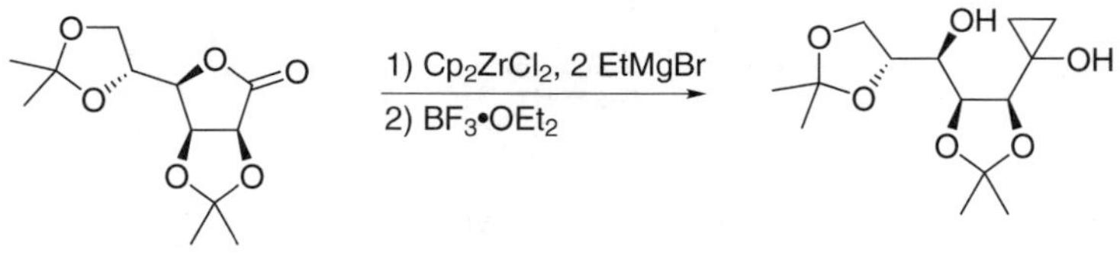

Scheme 5

Despite its undeniable synthetic utility, the Kulinkovich and related reactions cannot be applied to the synthesis of cyclopropanes from aldehydes and ketones for two reasons: (1) there is no leaving group to induce the ring contraction in this case (compare with Scheme 1); and (2) even at low temperature, Grignard reagents would rapidly react with carbonyl compounds. However, Szymoniak and coworkers have recently reported that vinylcyclopropanes could be prepared directly from α,β-unsaturated ketones (Scheme 6) [12]. In this method, the zirconacyclopropane intermediate **9** was preformed from Cp_2ZrCl_2 and the Grignard compound, before adding the carbonyl substrate. The reaction proceeds through the oxazirconacyclopentane **11** and ring contraction under the effect of a protic acid (Scheme 7).

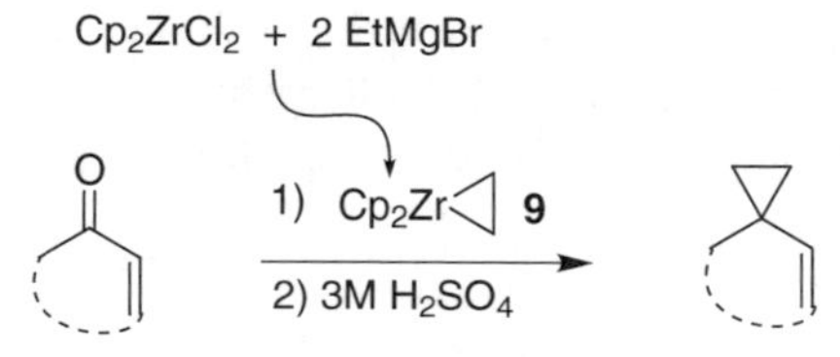

Zirconocene Complexes as New Reagents for the Synthesis of Cyclopropanes

Scheme 6

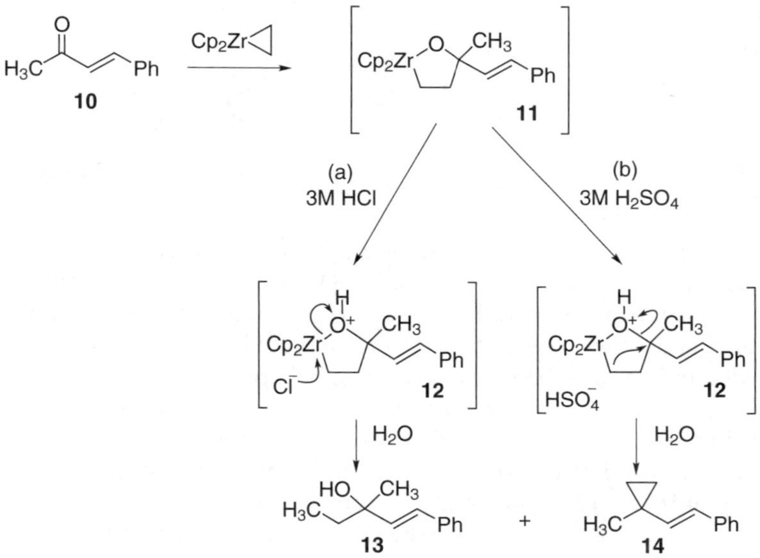

Scheme 7

[H⁺] = 3M HCl   78 : 22
[H⁺] = 3M H₂SO₄   8 : 92

Surprisingly, protonolysis of the reaction mixture with 3 M H₂SO₄ proved to be essential for the reaction to occur. The use of 3 M HCl or more dilute H₂SO₄ led to the predominant formation of the corresponding allylic alcohol **13** (paths b and a, respectively). A rationale for these two different reaction pathways has been proposed. In the first step, oxazirconacycle **11** is invariably protonated by HCl or H₂SO₄ to afford the oxonium intermediate **12**. When using HCl, cleavage of the Zr–O bond occurs under hydrolytic conditions with the assistance of the reasonably nucleophilic Cl⁻ (path a). In contrast, in the presence of weakly nucleophilic HSO₄⁻, competing ring contraction leads to the cyclopropane derivative **14** (path b).

The described reaction was applied to the synthesis of various vinylcyclopropanes. However, obvious limitations remained in the scope of the reaction at this stage: (1) only 1-substituted vinylcyclopropanes could be obtained; and (2) the reaction couldn't be applied to the preparation of cyclopropane derivatives other than vinylcyclopropanes. To overcome the first limitation, the methyl-substituted zirconacyclopropane [13] was stabilized by adding PMe₃ (complex **15**) prior to the reaction with the carbonyl compound (Scheme 8). Thus, the substituted cyclopropane **17** could be obtained in moderate yield. The use of a toxic and expensive additive, however, makes this approach of little synthetic value.

Scheme 8

The second limitation was solved by employing Lewis acid (TiCl$_4$) in a non-aqueous medium (CH$_2$Cl$_2$) instead of H$_2$SO$_4$ for the ring contraction [14]. In this way, cyclopropanes of various structure, including spiro compounds, were obtained in moderate to good yields from saturated, unsaturated, and aromatic aldehydes and ketones. Examples are given in Table 1 (entries 1–5).

The synthetic interest of the reaction is even broader since several functional groups in the substrate, such as C=C double bond, ether, halogen, ester, and amide groups (entries 6–10) can be present in the carbon skeleton. Chemoselective cyclopropanation reactions involved the enone functionality in the presence of a saturated ketone group (Scheme 9). Recently, the reaction was applied to the conversion of 1,3-dicarbonyl compounds into 2-alkoxyalkenyl cyclopropanes (Scheme 10) [15]. These molecules having both vinylcyclopropane and enol ether moieties are supposed to be versatile synthetic intermediates.

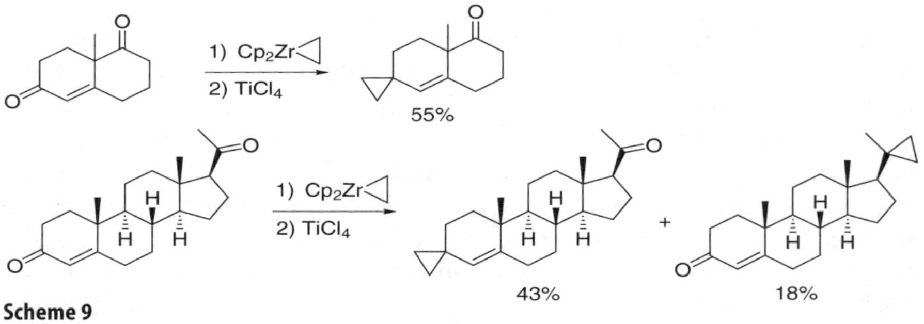

Scheme 9

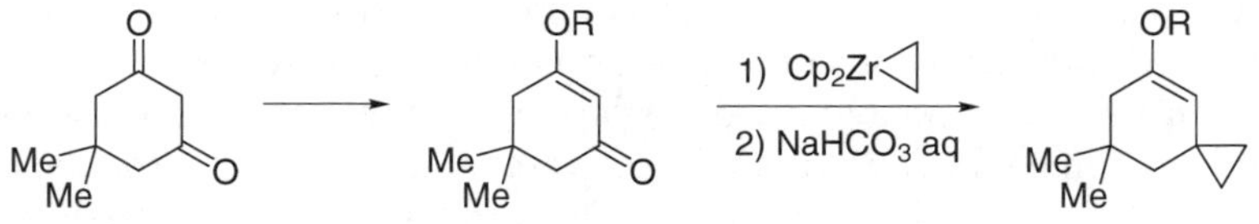

Scheme 10   R = Et, SiMe$_3$

Zirconocene Complexes as New Reagents for the Synthesis of Cyclopropanes 113

Table 1

$$R^1\underset{}{\overset{O}{\vphantom{X}}}R^2 \xrightarrow[\text{2) TiCl}_4]{\text{1) Cp}_2\text{ZrCl}_2,\ 2\ \text{EtMgBr}} R^1\underset{}{\overset{\triangle}{\vphantom{X}}}R^2$$

Entry	Substrate	Product	Yield (%)
1	n-C₇H₁₅–CH=CH–C(O)–Et	n-C₇H₁₅–CH=CH–C(cyclopropyl)–Et	60
2	4,4-dimethyl-6-methylcyclohex-2-enone	spiro-cyclopropane derivative	81
3	Ph–C(O)–CH=CH₂	Ph–C(cyclopropyl)–CH=CH₂	71
4	4-MeO-cyclohexadienyl cyclopropyl ketone	bis-cyclopropyl derivative	70
5	n-C₉H₁₉–C(O)–Me	n-C₉H₁₉–C(cyclopropyl)–Me	42
6	Me₂C=CH–CH₂–CH₂–C(Me)=CH–CHO	Me₂C=CH–CH₂–CH₂–C(Me)=CH–CH(cyclopropyl)	71
7	4-MeO–C₆H₄–CHO	4-MeO–C₆H₄–CH(cyclopropyl)–H	88
8	Ph–CH=C(Br)–CHO	Ph–CH=C(Br)–CH(cyclopropyl)–H	56
9	Et-substituted cyclohexenone with CO₂Et	spiro-cyclopropane derivative with CO₂Et	72
10	Ph–C(O)–CH₂–CH₂–C(O)–N(Me)–Bn	Ph–C(cyclopropyl)–CH₂–CH₂–C(O)–N(Me)–Bn	69

The synthesis of cyclopropanes from carbonyl compounds represents a new approach for converting zirconacycles into carbocycles via a deoxygenative ring contraction under Lewis acid activation. It differs in that from the spontaneous Kulinkovich process. Further progress would involve extension to the synthesis of 1,2-substituted cyclopropanes and the development of catalytic and enantioselective variants.

3
Construction of the Cyclopropane Ring from Three-Carbon Synthons

There are many synthetic approaches to cyclopropanes involving three-carbon synthons [16]. Among them, those employing metal reagents play an important role [17]. Zirconocene derivatives have recently been employed to construct the cyclopropane ring from a synthon having three carbons. The reported reactions can be divided into two groups as depicted in Scheme 11. In the first group, a γ-elimination of the X group takes place from the Zr(IV) intermediate to afford the cyclopropane compound whereas in the second group, the cyclopropanation step occurs through the intramolecular cyclization of a homoenolate anion equivalent.

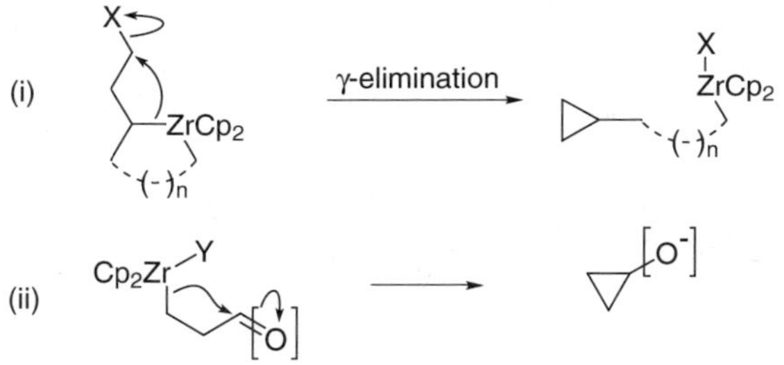

Scheme 11

3.1
Formation of Cyclopropanes via γ-Elimination Reactions

3.1.1
Zirconacycles as Intermediates

β-Elimination reactions are typically found in transition metal chemistry. The conversion of allylic ethers into allylzirconocenes [18] is a well-known zirconium-mediated process of this type. In contrast, the γ-elimination reactions are less frequently encountered, and only a few examples of γ-elimination involving zirconocenes have been reported. When studying isonitrile insertion into zirconacycles, Whitby and coworkers observed an interesting reaction leading to a cyclopropane derivative [19]. The overall transformation is depicted in Scheme 12. Treatment of the diene **18** with preformed butene–zirconocene gave

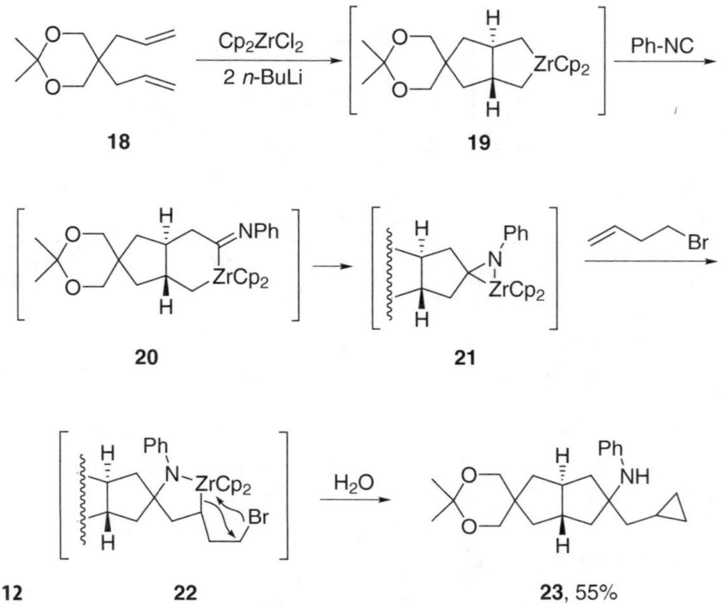

Scheme 12

the bicyclic zirconacyclopentane **19** [20]. The insertion of phenyl isocyanide gave the iminoacyl complex **20** that rearranged by warming to the η^2-imine complex **21**. The insertion of a homoallyl bromide into the Zr–C bond led to the azazirconacycle **22**. The following intramolecular transfer of the bromide to the metal (γ-elimination) and hydrolysis finally afforded the cyclopropane derivative **23**. The latter transformation seemed to represent a rather particular case, since C–Zr bonds do not normally react with alkyl halides.

Similarly, Takahashi and coworkers reported that treatment of alkynes with zirconocene–ethylene complex (**9**) and homoallylic bromides gave allylcyclopropane derivatives [21]. Therefore, the possibility of γ-elimination of the halogen atom in zirconacyclopentene intermediates appeared to be more general as expected. The plausible mechanism is shown in Scheme 13. Carbozirconation

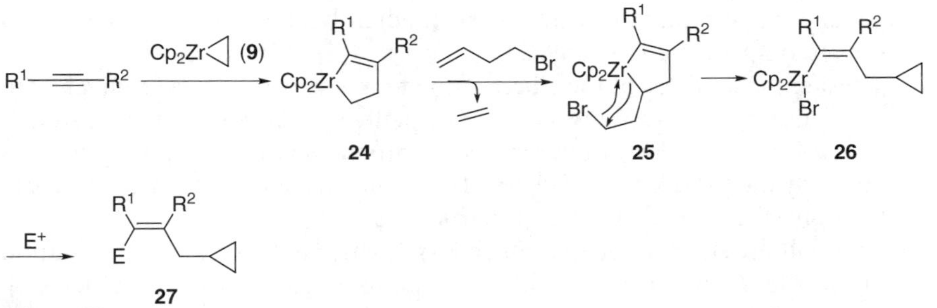

Scheme 13

Table 2

Entry	Alkyne	E+	Product	Yield (%)
1	n-Pr—≡—n-Pr	H+		68
2	n-Pr—≡—n-Pr	I₂		50
3	Ph—≡—Ph	H+		80
4	Ph—≡—Ph	D+		71
5	Ph—≡—n-C₅H₁₁	H+		68
6	Ph—≡—Me	H+	85 : 15	85
7	n-C₆H₁₃—≡—Me	H+		59

of alkynes with zirconocene–ethylene complex produces zirconacyclopentene derivatives **24**, which undergo an exchange reaction with the double bond of the homoallylic bromide to afford α-substituted zirconacyclopentene compounds **25**. Subsequent γ-elimination reaction produces **26**, which can be converted into the final cyclopropanes **27** by hydrolysis (deuterolysis) or iodinolysis. Selected examples are presented in Table 2.

The stereochemistry of the alkenyl moiety was very selective (equivalent to a *syn* addition). Interestingly, 2-bromo-4-pentene produced the *cis*-disubstituted cyclopropane derivative, whereas 1-bromo-2-ethyl-3-butene gave predominantly the *trans* isomer (Scheme 14). δ-Substituted homoallylic bromides did not give the corresponding cyclopropanes.

Homoallylic ethers appeared much less reactive towards the γ-elimination reaction with the zirconocene–alkyne complexes or zirconacyclopentenes [21]. No γ-elimination products were observed in these cases. In contrast, Szymo-

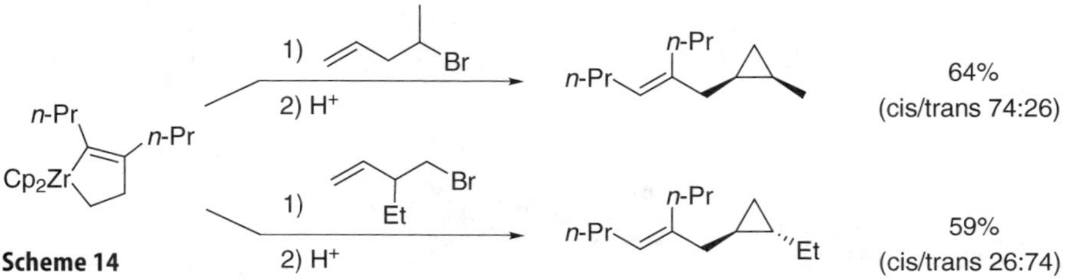

Scheme 14

[Scheme 14 showing reactions of Cp₂Zr(n-Pr)(n-Pr) alkene complex:
Top: 1) allyl bromide CH₂=CHCH(Me)Br, 2) H⁺ → n-Pr-CH=CH-cyclopropyl-n-Pr, 64% (cis/trans 74:26)
Bottom: 1) CH₂=C(Et)CH₂Br, 2) H⁺ → n-Pr-CH=CH-cyclopropyl(Et)(n-Pr), 59% (cis/trans 26:74)]

niak and coworkers have recently demonstrated that benzylic homoallylic ethers may easily be converted into cyclopropane derivatives when zirconacyclopropanes are involved as intermediates [22]. The reaction of these homoallylic ethers with zirconocene complexes (Cp_2ZrCl_2/2 n-BuLi) proceeds through γ-elimination of the alkoxy group and leads to the formation of cyclopropylcarbinylzirconium complexes in mild conditions. The hydrolysis of the reaction mixture affords the corresponding cyclopropanes (Scheme 15).

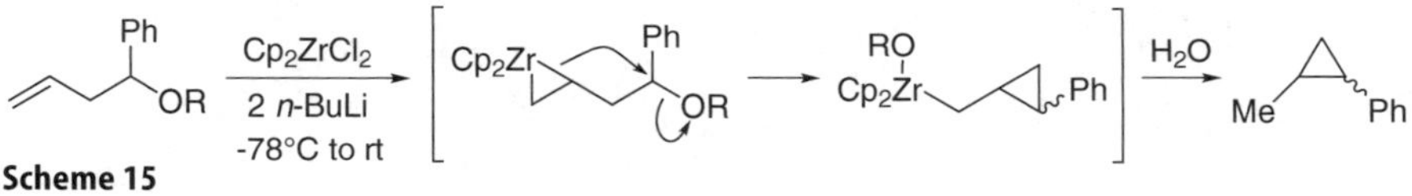

Scheme 15

[Scheme 15: PhCH(OR)CH₂CH=CH₂ + Cp₂ZrCl₂/2 n-BuLi, −78 °C to rt → zirconacyclopentane intermediate with Ph and OR → cyclopropylcarbinyl-Cp₂Zr(OR)-Ph → H₂O → Me-cyclopropyl-Ph]

In several cases, cyclopropylcarbinyl–homoallyl rearrangements involving zirconium species were observed (Scheme 16), similar to those reported for some other transition metals [23]. Examples of reactions are depicted in Table 3. The lack of rearrangement for the homoallylic ethers in entries 3–5 was attributed to the Thorpe–Ingold effect, i.e., angle compression at the substituted carbon [24].

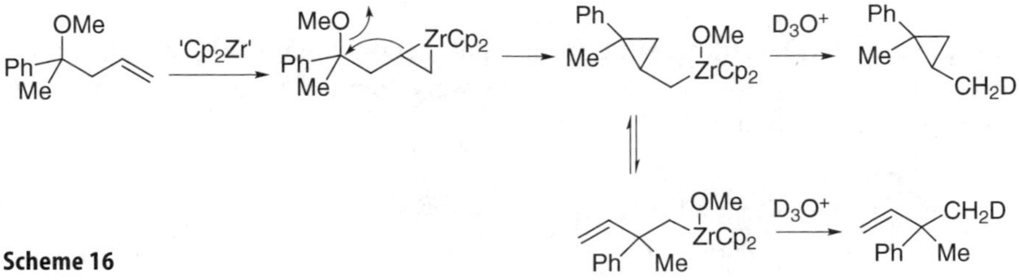

Scheme 16

[Scheme 16: PhCH(Me)CH(OMe)CH₂CH=CH₂ → 'Cp₂Zr' → zirconacyclopentane with Ph, Me, OMe → Ph-cyclopropyl(Me)(OMe)ZrCp₂ → D₃O⁺ → Ph-cyclopropyl(Me)(CH₂D); equilibrium with homoallyl form PhCH=CHCH(Me)CH(OMe)ZrCp₂ → D₃O⁺ → PhCH=CHCH(Me)CH₂D with Me]

3.1.2
Reactions Initiated by Hydrozirconation

The γ-elimination approaches to cyclopropanes [Scheme 11, Eq. (i)] would be a quite general method, regardless of the way in which the precursor zirconium intermediate was formed. The hydrozirconation reaction [25] can be considered in this respect, since it provides an easy and frequently used route to organozirconocenes. The residual formation of cyclopropane rings through the hydrozirconation–γ-elimination sequence had already been observed in 1976

Table 3

Entry	Substrate	Products		Yield (%) (trans/cis)
1	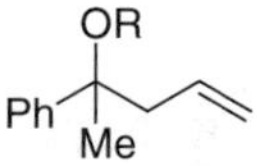 Ph-C(OR)(Me)-CH2-CH=CH2	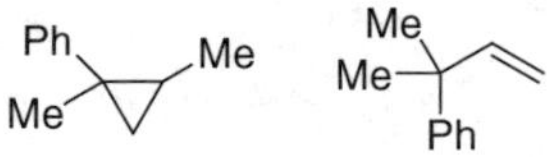 Ph/Me cyclopropane + Me2C(Ph)-CH=CH2		
	R = Me	7.8 : 1		85 (2.8 : 1)
	R = Bn	6.3 : 1		50 (3.2 : 1)
	R = MOM	7.4 : 1		60 (3.0 : 1)
2	Ph-C(OMe)(n-Bu)-CH2-CH=CH2	Ph/n-Bu cyclopropane + Me2C(Ph)-CH=CH2	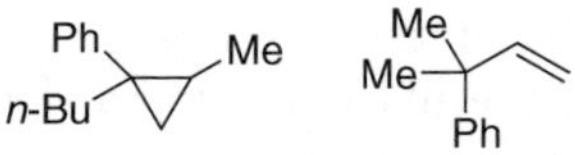 14.5 : 1	79 (1.1 : 1)
3	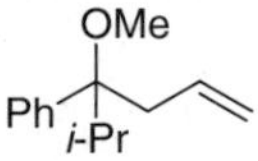 Ph-C(OMe)(i-Pr)-CH2-CH=CH2	 Ph/i-Pr cyclopropane with Me		80 (1.4 : 1)
4	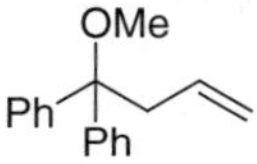 Ph-C(OMe)(Ph)-CH2-CH=CH2	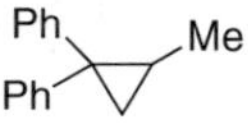 Ph/Ph cyclopropane with Me		80
5	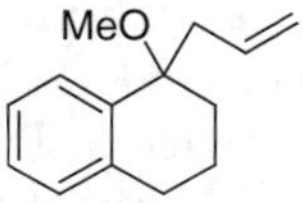 1-MeO-1-allyl-tetralin	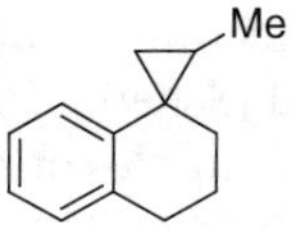 spirocyclopropane-tetralin with Me		76 (1 : 1)

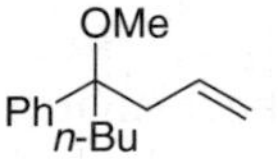

by Tam and Rettig [26]. Starting from 4-halogenoalkanes (X=Cl, Br), mixtures of cyclopropanes, alkenes, and alkanes were formed in these reactions, as exemplified in Scheme 17.

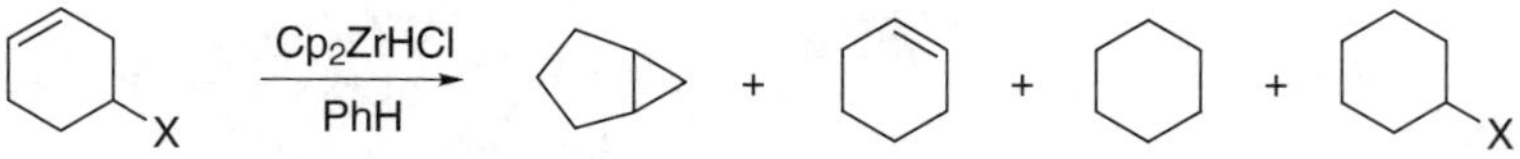

4-halocyclohexene →(Cp2ZrHCl, PhH)→ bicyclo[3.1.0] + cyclohexene + cyclohexane + halocyclohexane

X = Cl	40%	26%	5%	15%
X = Br	18%	40%	4%	26%

Scheme 17

In 1997, Taguchi, Hanzawa and coworkers described the formation of cyclopropyl carbinols from alkenyl oxiranes through chemoselective hydrozirconation followed by intramolecular attack of the generated alkylzirconocene on the epoxide ring (Scheme 18) [27]. Both the reactivity and the selectivity patterns are worthy of note. The sequence was made possible owing to the chemoselective hydrozirconation of the alkene in the presence of oxirane. Despite the poor nu-

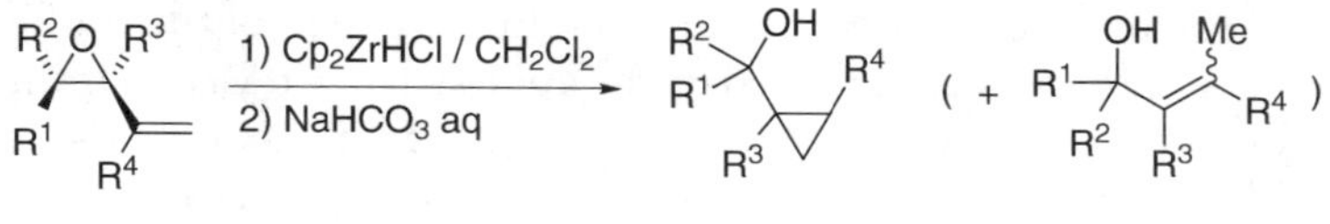

Zirconocene Complexes as New Reagents for the Synthesis of Cyclopropanes 119

Scheme 18

cleophilicity of alkylzirconocenes, intramolecular nucleophilic cyclization took place. The ring opening of the epoxide provides the driving force for the intramolecular cyclization of the poor nucleophilic zirconium species. The concomitant formation of allylic alcohols **30** was considered to be a result of the regioisomeric hydrozirconation in the presence of the chelating oxygen atom [25a], followed by β-elimination. Examples of these reactions are depicted in Table 4.

Table 4

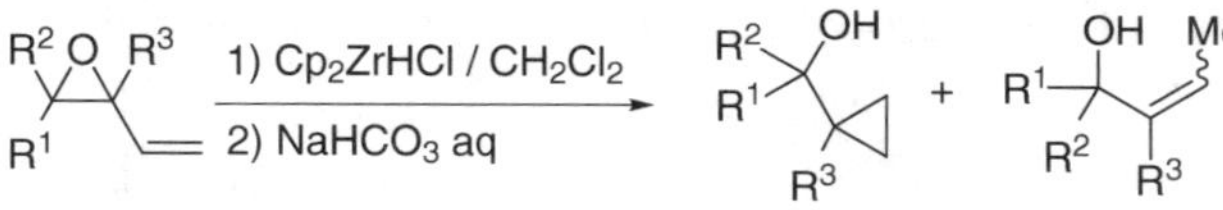

Entry	Substrate	Products
1	 Bn-epoxide-vinyl	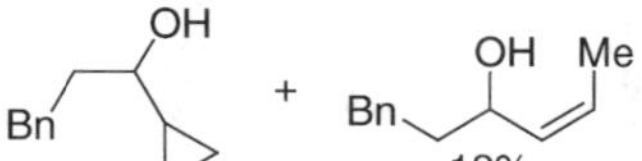 Bn-CH(OH)-cyclopropyl 72% + Bn-CH(OH)-CH=CH-Me 18%
2	Bn-epoxide-vinyl (trans)	Bn-CH(OH)-cyclopropyl 59% + Bn-CH(OH)-CH=CH-Me 15%
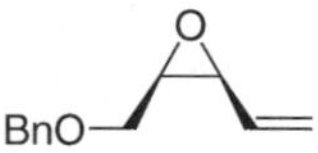 3	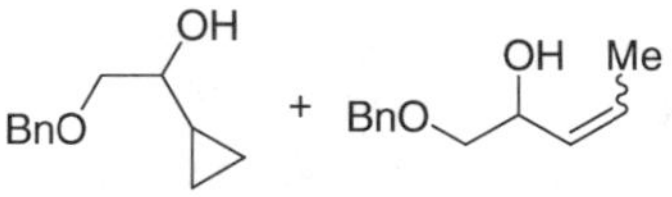 BnO-epoxide-vinyl	BnO-CH(OH)-cyclopropyl 52% + BnO-CH(OH)-CH=CH-Me 22% (E/Z = 1:7.3)
4	Bn-epoxide(Me)-vinyl	 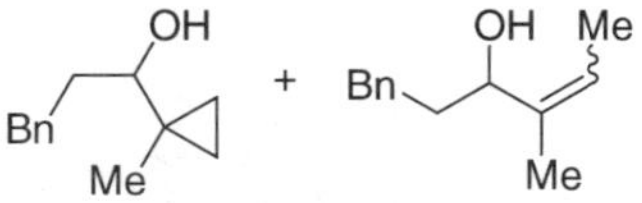 Bn-CH(OH)-cyclopropyl(Me) 76% + Bn-CH(OH)-C(Me)=CH-Me 2% (E/Z = 1:1.9)
5	Ph-epoxide-vinyl	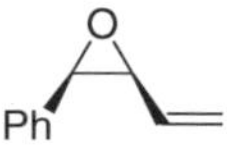 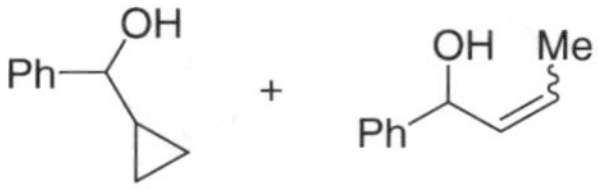 Ph-CH(OH)-cyclopropyl 70% + Ph-CH(OH)-CH=CH-Me 5% (E/Z = 1:5.1)

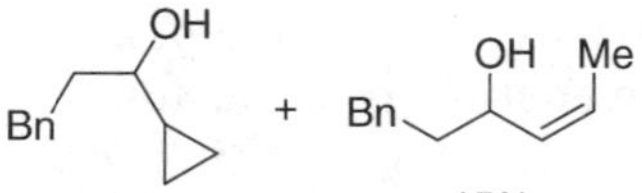

The stereochemical outcome of the reaction (inversion of the configuration at the oxirane carbon) was explained by the approach of Cp$_2$ZrHCl from the less hindered side of the stable *gauche* conformer of the vinyloxirane (Scheme 19). *Cis*-epoxides gave only *syn-trans* cyclopropyl carbinols, whereas *trans*-oxiranes gave mixtures of *anti-trans* and *anti-cis* isomers.

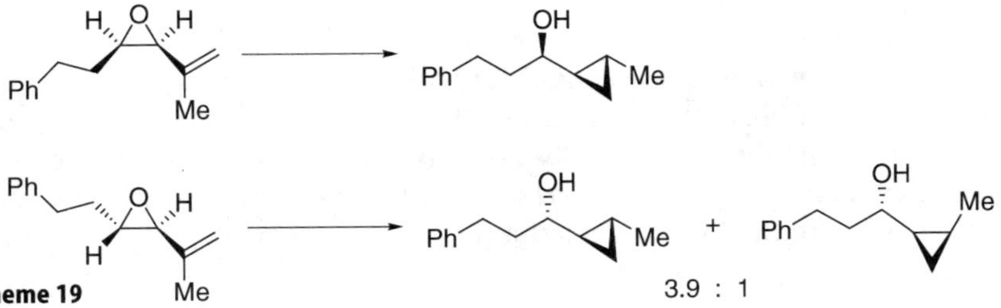

Scheme 19

Similarly to vinyloxiranes, the conversion of vinylaziridine derivatives **31** with Cp$_2$ZrHCl to (cyclopropylmethyl)amines **32** was also performed in good yields (Scheme 20).

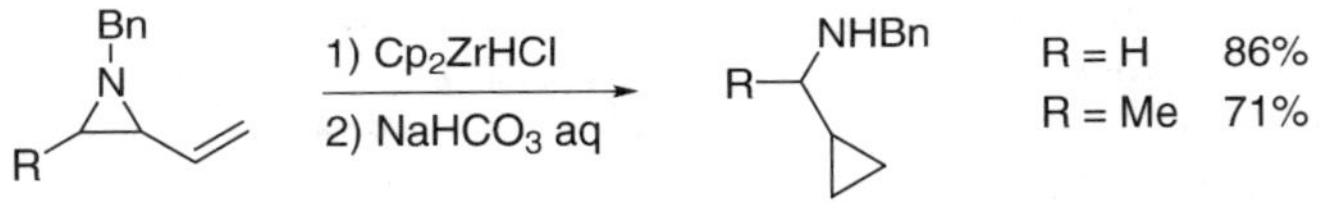

Scheme 20

A more general approach to cyclopropanes from allylic ethers has recently been reported [28]. Two steps are involved in this one-pot transformation, i.e., hydrozirconation reaction followed by Lewis acid-promoted deoxygenative ring formation. The overall transformation is symbolized in Scheme 21, Eq. 1,

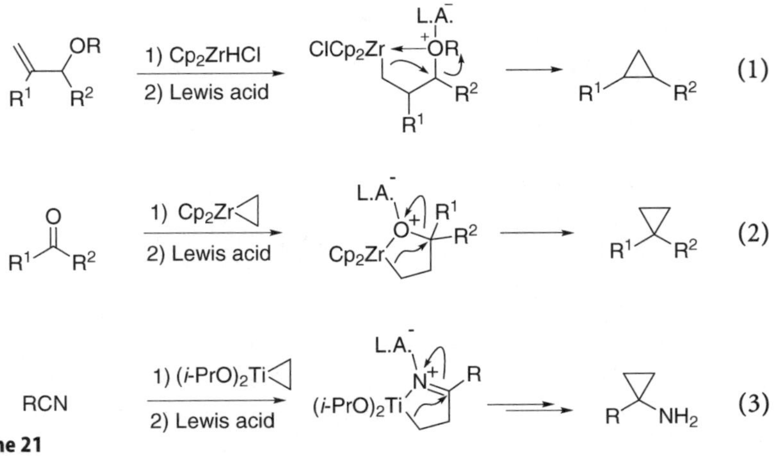

Scheme 21

Table 5

Entry	Substrate	Product	Yield (%)
1	Ph, OMe-substituted allyl ether	Ph-CH₂-cyclopropyl	80
2	2-naphthyl, OBn allyl ether	2-naphthyl-CH₂-cyclopropyl	89
3	2-naphthyl, OBn, Me quaternary allyl ether	2-naphthyl-C(Me)-cyclopropyl	79
4	Me, OBn prenyl-type allylic ether	Me-substituted cyclopropyl derivative	78
5	2-vinyl-1,3-dioxolane	HO-CH₂CH₂-O-cyclopropyl	65
6	CH₂=CH-CH(OEt)₂	EtO-cyclopropyl	92

Reaction conditions: Cp₂Zr(H)Cl (1.0 eq.), BF₃·OEt₂ (1.1 eq.), CH₂Cl₂, rt

and represents a process analogous to the previously described synthesis of cyclopropanes from carbonyl compounds, Eq. 2 [14], and nitriles, Eq. 3 [7]. The reactions employed methyl or benzyl ethers bearing both alkyl and aryl substituents (Table 5). Quaternary centers as well as substituted double bonds can be present in the substrate (entries 3 and 4). The reactions employing allylic acetals can be carried out, leading to cyclopropyl ethers (entries 5 and 6).

2-Substituted allylic ethers were transformed predominantly into *trans*-1,2-di- and -1,1,2-trisubstituted cyclopropanes (Table 6). The degree of *trans* selectivity depends on the OR group, and is higher with R=Me than with Bn (entries 1 and 2), the diastereoselectivity being generated in the hydrozirconation step [29]. Furthermore, the cyclization step proceeds with the inversion of the configuration at the carbon atom bound to zirconium [30].

Table 6

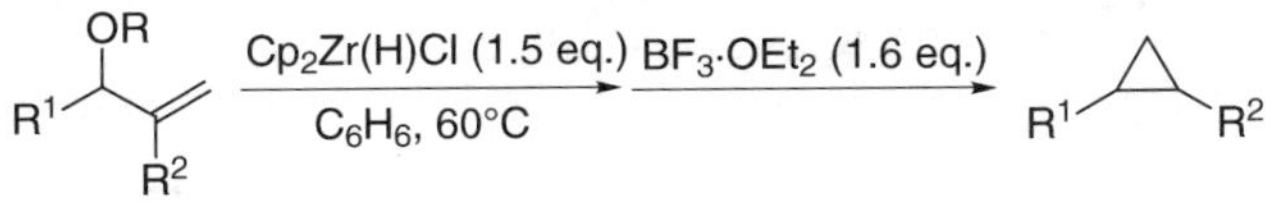

$$R^1 \underset{R^2}{\overset{OR}{\diagup\!\!\!\!\diagdown}} \xrightarrow[C_6H_6,\ 60°C]{Cp_2Zr(H)Cl\ (1.5\ eq.)\ BF_3 \cdot OEt_2\ (1.6\ eq.)} R^1 \triangleleft R^2$$

Entry	Substrate	Product	Yield (%) (trans:cis)
1	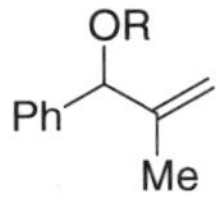 Ph–C(OR)(Me)–CH=CH₂	R = Bn R = MOM R = Me	Ph–△–Me	76 (71 : 29) 80 (85 : 15) 86 (97 : 3)
2	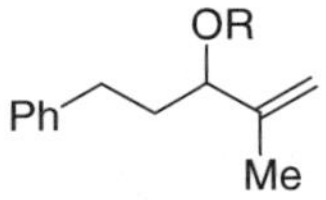 Ph–CH₂–CH₂–C(OR)(Me)=CH₂	R = Bn R = Me	Ph–CH₂–CH₂–△–Me	75 (83 : 17) 72 (90 : 10)
3	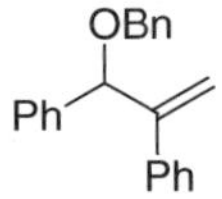 Ph–CH(OBn)–C(Ph)=CH₂	 Ph–△–Ph	60 (100:0)
4	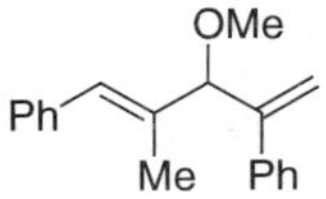 Ph–CH=C(Me)–CH(OMe)–C(Ph)=CH₂	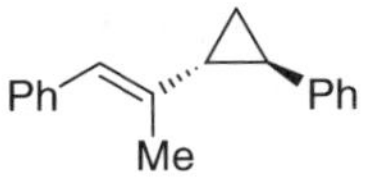 Ph–CH=C(Me)–△–Ph	75 (90:10)
5	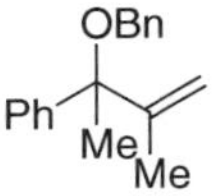 Ph–C(OBn)(Me)–C(Me)=CH₂	 Ph(Me)–△–H(Me)	42 (100:0)

An early report of a cyclopropane-forming reaction involves the initial zirconation of a diene (Scheme 22) [31]. The cyclopropanation step took place from the unsaturated zirconocene intermediate and NBS. The reaction has not been presented in terms of the γ-elimination process, but the real mechanism remains unclear.

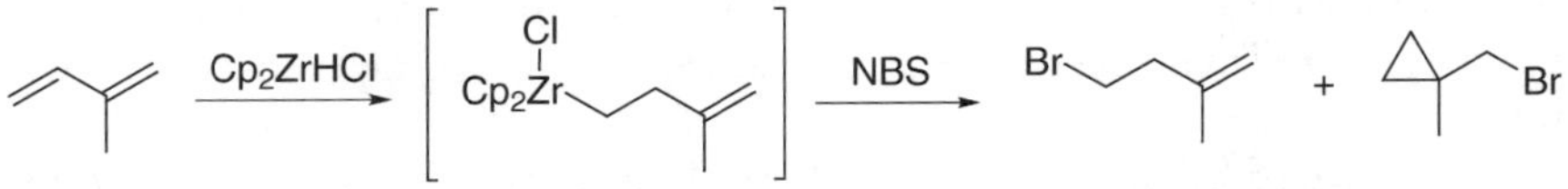

$$\text{CH}_2\!=\!\text{C(Me)}\!-\!\text{CH}\!=\!\text{CH}_2 \xrightarrow{Cp_2ZrHCl} \left[Cp_2Zr(Cl)\text{-CH}_2\text{CH}_2\text{-C(Me)}\!=\!\text{CH}_2 \right] \xrightarrow{NBS} \text{Br-CH}_2\text{-C(Me)}\!=\!\text{CH}_2 + \triangleleft\text{-CH}_2\text{Br}$$

Scheme 22   85% (56 : 44)

3.2
Cyclopropanes from Homoenolate Anion Equivalents

The title transformation provides an alternative pathway for the construction of the cyclopropane ring from a synthon of three carbons. The intramolecular cyclization of homoenolate anion equivalents is represented in Scheme 11, Eq. (ii). In practice, several routes involving metals such as Mg, Zn, and Ti, as well as different homoenolate precursors, have been developed [17a–e].

An interesting approach to cyclopropanes involving zirconocene chemistry has recently been described by Taguchi and coworkers [32]. In this method, γ,γ-dialkoxyzirconium species **35** were first generated in situ by the reaction of orthoacrylic acid triethyl ester **33** with zirconocene–butene complex (Scheme 23). The reaction proceeds through the addition–β-elimination sequence, similar to the conversion of allylic ethers into allylzirconocenes [18]. The complex **35** revealed chameleonic reactivity, depending on the reaction conditions. It proved to possess two reactive sites, one being the γ-carbon (typical reactivity of allylic organometallics) [33] and the second being the β-carbon (reactivity of a ketene dialkyl acetal) [32]. Thus, the complex reacted selectively at the γ-position in the absence of a Lewis acid [33], or in the presence of 0.2–0.3 equiv of BF$_3$·OEt$_2$ [34], leading to 1-substituted-2,2-diethoxy-2-buten-1-ol derivatives **36**. In contrast, in the presence of more than 1 equiv of BF$_3$·OEt$_2$ or Me$_3$SiOTf [32], **35** reacted first at the β-position followed by a cyclization reaction to afford the cyclopropanone acetals **37** (Scheme 24). Obviously, under these conditions, the reactivity of ketene dialkyl acetal overcomes that of the allylic zirconium.

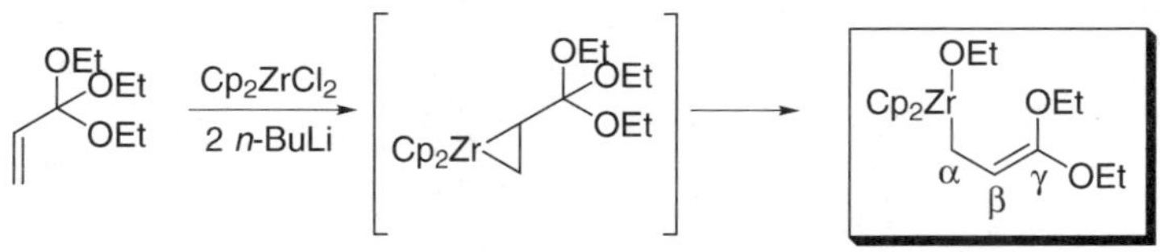

Scheme 23

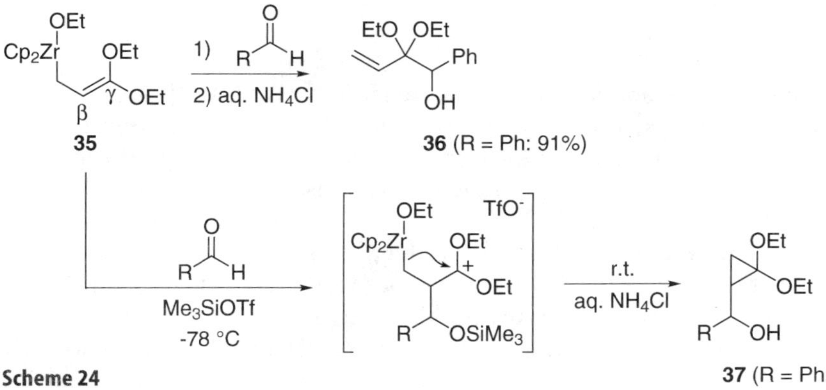

Scheme 24

Table 7

Entry	Carbonyl compound	Product	Yield (%)
1	Ph~~CHO	Ph-CH(OH)-CH2-C(OEt)2 cyclopropane	88
2	CyclohexylCHO	Cyclohexyl-CH(OH)-C(OEt)2 cyclopropane	87
3	n-C7H15CHO	n-C7H15-CH(OH)-C(OEt)2 cyclopropane	75
4	PhCHO	Ph-CH(OH)-C(OEt)2 cyclopropane	81
5	cyclohexanone	cyclohexane-C(OH)-C(OEt)2 cyclopropane spiro	83
6	Ph~~C(O)~~CH=CH2	Ph-CH2CH2-C(O)-CH2-C(OEt)2 cyclopropane + Ph-CH2CH2-C(OH)(=CH2)-C(OEt)2 cyclopropane	78 + 15
7	cyclohex-2-enone	3-(2,2-diethoxycyclopropyl)cyclohexanone	96

Representative examples are summarized in Table 7. Aliphatic and aromatic aldehydes as well as ketones could be used (entries 1–5). α,β-Unsaturated substrates gave predominantly or exclusively 1,4-addition compounds (entries 6 and 7). Under the Lewis acid-promoted conditions, toluene was found to be a more effective solvent than THF.

Another site-selective reaction of **35** with α,β-unsaturated carboxylic acid derivatives in the presence of a Lewis acid has been reported (Scheme 25) [35]. The intermediate **37** possesses two nucleophilic sites: the zirconium alkyl and the enol ether moieties. Starting from **37**, two competing pathways lead accordingly to cyclopropane (**38**, path a) and/or cyclobutane (**39**, path b) derivatives. The

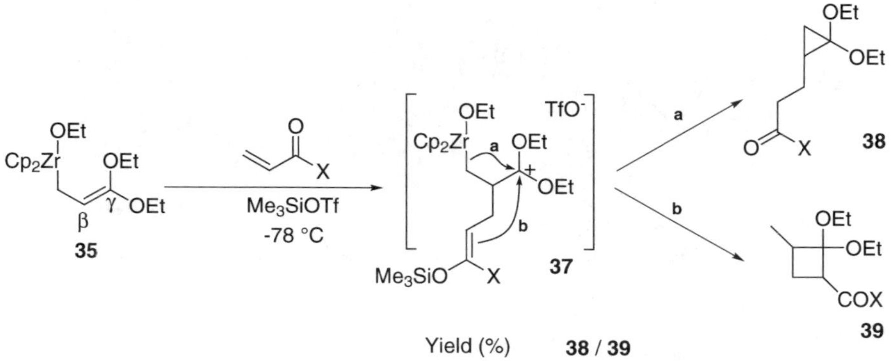

Scheme 25

	Yield (%)	38 / 39
X = OBn	32	22 : 78
X = NMe₂	54	<5 : 95
X = NBn₂	83	<5 : 95
X = NPh₂	61	75 : 25

cyclobutane–cyclopropane ratio can be controlled by the electron density at the COX group. When using N,N-diphenylacrylamine, a cyclopropane derivative was obtained as the major product. A total cyclopropane vs cyclobutane selectivity was achieved by the use of α,β-unsaturated N-acyloxazolidinones; in these cases, the use of CH_2Cl_2 as solvent and Et_2AlCl as Lewis acid were required for the reaction to occur [36].

4
Miscellaneous Cyclopropane Syntheses

This section is especially devoted to syntheses in which the zirconocene complexes are involved initially, but not directly in the cyclopropanation step. Typically, the transmetalation to other metals such as Al, Zn or Cu proceeds at first. Some other cases can also be noticed. Among them, an early report by Negishi represents examples of migratory insertion reactions of organylzirconocene chlorides with α-haloorganolithiums 40 [37]. In this way, trisubstituted cyclopropanes were obtained as shown in Scheme 26. High stereoselectivity was observed in these reactions, leading almost exclusively to E-configured cyclopropanes.

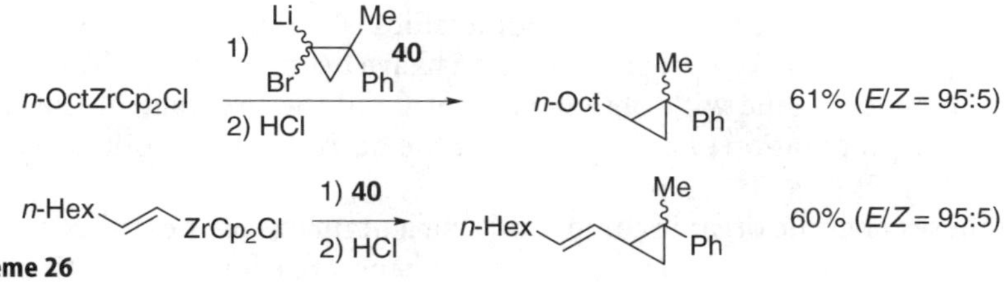

Scheme 26

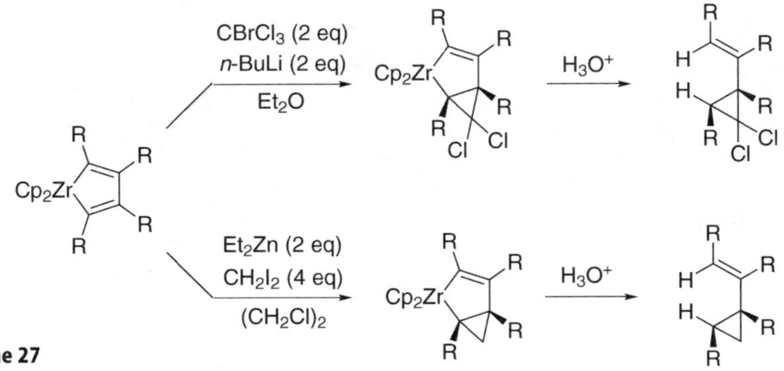

Scheme 27

Examples of a rather classical cyclopropanation reaction involving Simmons–Smith-type reagents or dichlorocarbene and a double bond of zirconacyclopentadienes were reported by Takahashi and coworkers (Scheme 27) [38].

Interestingly, carbonylation of the intermediate **41** afforded substituted benzene derivative **45** (Scheme 28). The mechanism proposed for this reaction involves CO insertion into the Csp^3–Zr bond of **41** to afford zirconacyclohexenone **42**. The latter undergoes intramolecular nucleophilic addition of the Csp^2–Zr bond to a carbonyl group to give **43**, which in turn rearranges with deoxygenation leading to **44** and **45**.

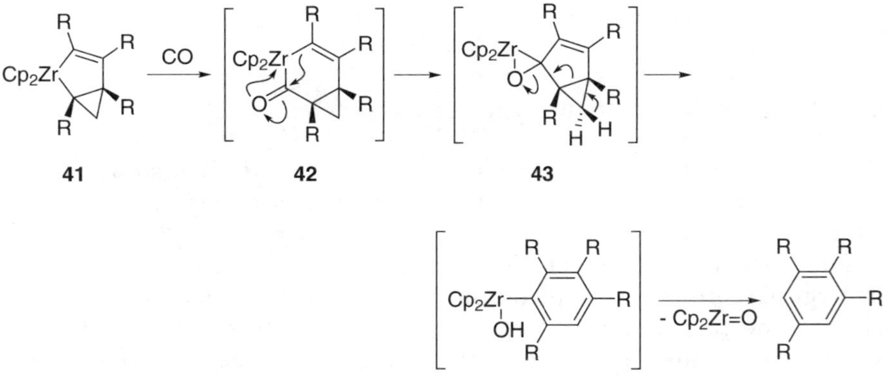

Scheme 28

The synthetic potential of organozirconocenes is greatly expanded by their easy transformation into other organometallics. Therefore, transmetalation-based approaches to cyclopropane synthesis have been reported. The reaction of zirconacyclopentene with phthaloyl chloride in the presence of CuCl was used for the preparation of cyclopropylenolate derivatives in moderate yields (Scheme 29) [39].

Syntheses of cyclic organic compounds via aluminacycles, described mainly by the Dzhemilev group, constitute a promising area [40–43]. Particularly,

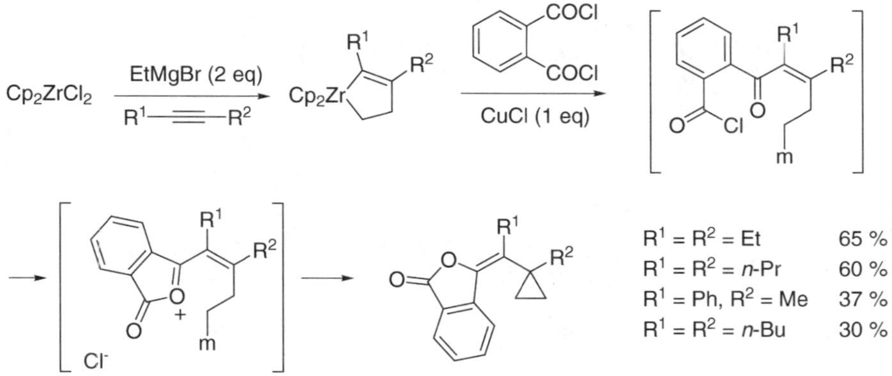

Scheme 29

R¹ = R² = Et   65 %
R¹ = R² = n-Pr 60 %
R¹ = Ph, R² = Me 37 %
R¹ = R² = n-Bu 30 %

aluminacyclopentanes or -pentenes formed in situ were converted into cyclopropane derivatives in different ways. The reaction of 3-alkyl-substituted aluminacyclopentanes with an excess of allyl chloride (reoxidant) in the presence of a catalytic amount of Ni(acac)₂ was described (Scheme 30) [44, 45]. The replacement of allyl chloride by crotyl halides (X=Cl, Br) led to the formation of regioisomeric cyclopropanes (Scheme 31) [45]. The mechanism of these reactions was proposed.

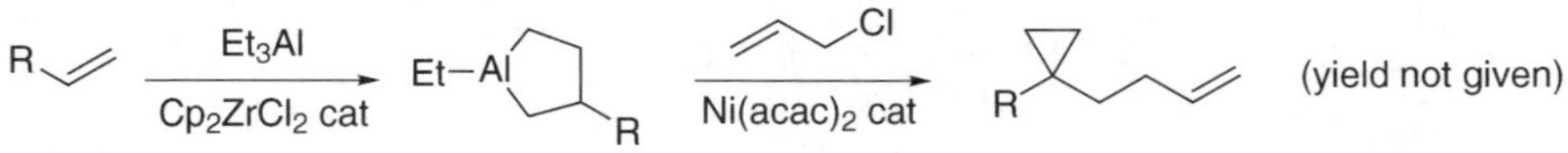

Scheme 30

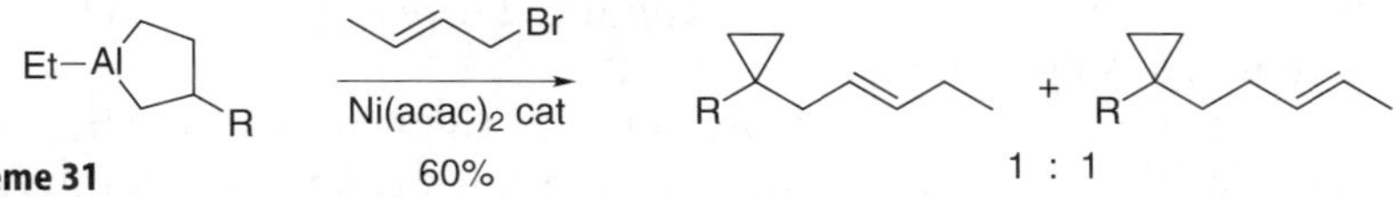

Scheme 31

The method was next applied to the synthesis of cyclopropanes from 2,3-substituted aluminacyclopentenes (Scheme 32) [45, 46]. In this case, a regioselective cleavage of the Csp²–Al bond of the intermediate aluminacycle **46** was achieved by using dialkyl sulfates as alkylating agents. The resulting open intermediate **47** undergoes homoallyl–cyclopropylcarbinyl rearrangement (intramolecular carboalumination) to afford cyclopropylcarbinylaluminum species **48**, that upon additional alkylation is converted into 1,1-disubstituted cyclopropanes **49**.

A synthetic route to alkenylcyclopropanes including bicyclic compounds via zirconium-catalyzed carboalumination of alkynes and enynes was reported by Negishi and coworkers [47]. The reaction sequence is shown in Scheme 33. Starting from the intermediate aluminacycle **50**, the reaction proceeds through a selective ring opening with XCH₂OMe (X=Cl, Br), followed by homoallyl–

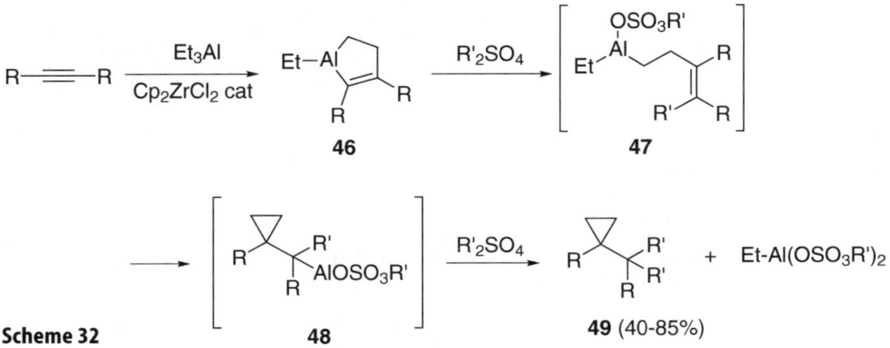

Scheme 32

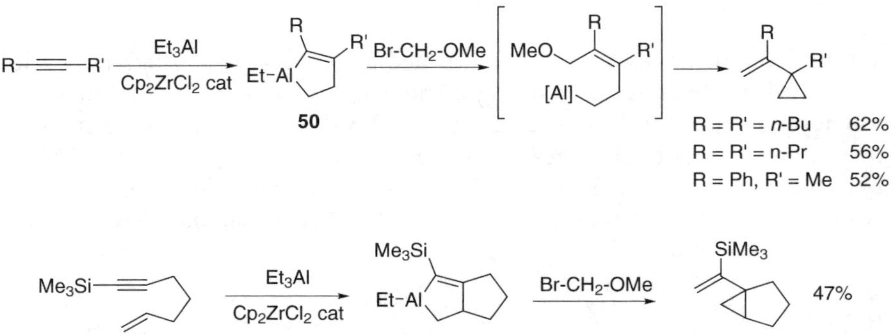

Scheme 33

cyclopropylcarbinyl rearrangement and deoxygenation. Zirconium-catalyzed carboalumination possibly offers opportunities for performing enantioselective reactions by using chiral zirconocenes.

Transmetalation reactions from zirconium to zinc have recently been employed by Wipf and coworkers to develop a versatile methodology for the synthesis of aminocyclopropane derivatives [48, 49]. The reaction constitutes a three-component coupling of in situ prepared alkenyl zirconocenes, aldimines, and dihalomethanes (Scheme 34). When the reaction was carried out in THF allylic amines **54** were formed, whereas in CH_2Cl_2 the *trans*-configured C-cyclopropylalkylamines **55** were obtained. The proposed mechanism for cyclopropane formation involves hydrozirconation of alkyne and in situ transmetalation to Me_2Zn, followed by addition of the phosphinoylamine to give allylic amide **51**. The successive reaction with CH_2Cl_2 leads to the zinc carbenoid species **52**, which serves in a Simmons–Smith-type transformation to afford the cyclopropane derivative **53**. The effective use of CH_2Cl_2 for cyclopropanation in the Simmons–Smith reaction is unprecedented. The cyclopropane products were obtained as well by using CH_2Cl_2 or CH_2I_2 in toluene solution. The use of enynes as substrates provides conjugated biscyclopropanes through the stereoselective formation of five C–C bonds (Scheme 35) [50].

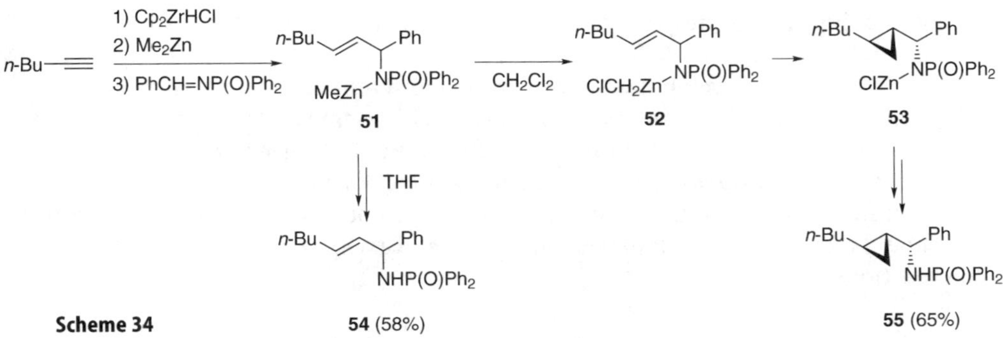

Scheme 34

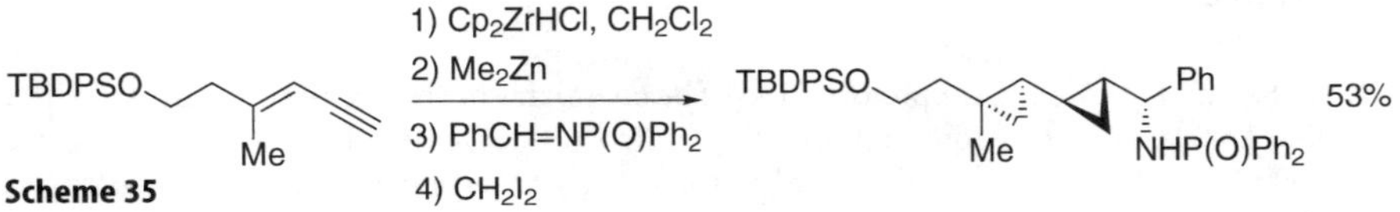

54 (58%)   55 (65%)

Scheme 35

53%

The cyclopropanation reactions described by Wipf and coworkers provide an interesting case of the bimetallic (here Zr–Zn) combination displaying unique reactivity [51, 52]. Actually, in addition to zinc species, the presence of zirconocene appeared essential for the cyclopropanation to occur. As depicted in Scheme 36, the mechanistic proposal is based on the dual Lewis acid and halide bridging properties of the zirconocene part [50].

Scheme 36

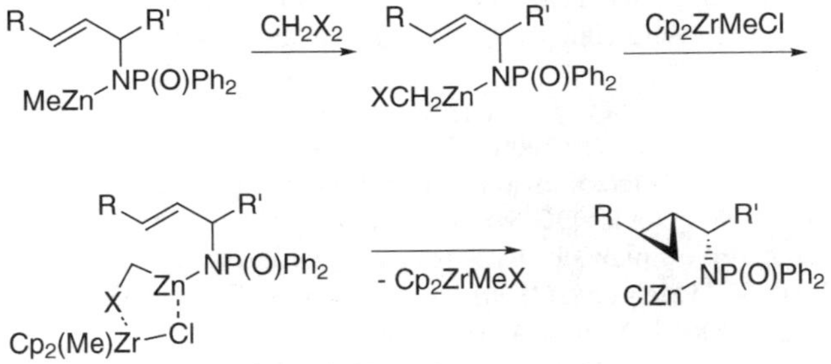

5
Conclusion

By using zirconocene complexes, cyclopropanes can be formed in various ways from diverse starting materials. The deoxygenative processes involving two- and one-carbon building blocks offer interesting alternatives to the more classical methods that employ alkenes and carbenoid species. The γ-elimination process through three-carbon zirconocene intermediates constitutes a basis for several reactions. Some of them exhibit subtle selectivity patterns, and could be synthetically useful. Transmetalation-based reactions give opportunities for developing new routes to cyclopropanes by using the chemistry of various metals.

However, the development of cyclopropane synthesis through zirconocene chemistry is still in its infancy. The reactions presented in this chapter have only recently been reported for the most part, and not systematically studied. Further investigations appear to be desirable. Practical procedures involving optimized reaction conditions and simpler reagents would be welcomed. Advances should focus on the development of catalytic and asymmetric cyclopropanation reactions.

References

1. (a) Salaün J (1987) In: Rappoport Z (ed) The chemistry of the cyclopropyl group. Wiley, New York, p 809; (b) Paquette LA (1991) In: Trost BM, Fleming I (eds) Comprehensive organic synthesis. Pergamon, Oxford, vol 5, p 899; (c) Murphy PJ (1995) In: Katrisky AR, Meth-Cohn O, Rees CW (eds) Comprehensive organic functional group transformations. Pergamon, Oxford, vol 1, p 801; (d) de Meijere A (1997) Methods of organic chemistry, vol E17a-f. Houben-Weyl, Thieme, Stuttgart
2. (a) Liu HW, Walsh CT (1987) In: Rappoport Z (ed) The chemistry of the cyclopropyl group. Wiley, New York, p 959; (b) Suckling CJ (1988) Angew Chem Int Ed Engl 27:537; (c) Salaün J, Baird MS (1995) Curr Med Chem 2:511; (d) Salaün J (2000) Top Curr Chem 207:1
3. (a) Donaldson WA (2001) Tetrahedron 57:8589; (b) de Meijere A (2003) Chem Rev 103:931
4. (a) Pfaltz A (1998) In: Beller M, Bohm C (eds) Transition metals for organic synthesis, vol 1. Wiley-VCH, Weinheim, p 100; (b) Lebel H, Marcoux JF, Molinaro C, Charette AB (2003) Chem Rev 103:977
5. (a) de Meijere A, Kulinkovich OG (2000) Chem Rev 100:2789; (b) Sato F, Urabe H, Okamoto S (2000) Chem Rev 100:2835; (c) de Meijere A, Koshushkov SI, Savchenko AI (2002) In: Marek I (ed) Titanium and zirconium in organic synthesis. Wiley-VCH, Weinheim, p 390; (d) Kulinkovich OG, Sviridov SV, Vasilevsky DA, Prityckaja TS (1989) Zh Org Khim 25:2244; (e) Kulinkovich OG, Sviridov SV, Vasilevsky DA (1991) Synthesis 234
6. (a) Chaplinski V, de Meijere A (1996) Angew Chem Int Ed Engl 35:413; (b) Chaplinski V, Winsel H, Kordes M, de Meijere A (1997) Synlett 111
7. (a) Bertus P, Szymoniak J (2001) Chem Commun 1792; (b) Bertus P, Szymoniak J (2002) J Org Chem 67:3965; (c) Bertus P, Szymoniak J (2003) Synlett 265; (d) Laroche C, Bertus P, Szymoniak J (2003) Tetrahedron Lett 44:2485; (e) Bertus P, Szymoniak J (2003) J Org Chem 68:7133
8. (a) Negishi E, Takahashi T (1994) Acc Chem Res 27:124; (b) Negishi E, Takahashi T (1998) Bull Chem Soc Jpn 71:755
9. (a) Burdon J, Price RC (1986) J Chem Soc Chem Commun 893; (b) Elphimoff I, Sarda P (1975) Tetrahedron 31:2785; (c) Motherwell WB, Roberts LR (1992) J Chem Soc Chem Commun 1582; (d) Harrison IT, Rawson RJ, Turnbull P, Fried JH (1971) J Org Chem 36:3515; (e) Vargas RM, Theys RD, Hossain MM (1992) J Am Chem Soc 114:777; (f) Csuk R, Höring U, Schaade M (1996) Tetrahedron 29:9759; (g) Imamoto T, Kamiya Y, Hatajima T, Takahashi H (1989) Tetrahedron Lett 30:5149; (h) Höppe HA, Lloyd-Jones GC, Murray M, Peakman TM, Walsh KE (1998) Angew Chem Int Ed 37:1545

10. Fujita K, Yorimitsu H, Shinokubo H, Matsubara S, Oshima K (2001) J Am Chem Soc 123:12115
11. V. Gandon (2002) PhD thesis, Université de Reims
12. Bertus P, Gandon V, Szymoniak J (2000) Chem Commun 171
13. Substituted zirconacyclopropanes were demonstrated to be less stable than zirconacyclopropane itself, see [8]
14. Gandon V, Bertus P, Szymoniak J (2000) Eur J Org Chem 3713
15. Sayouri F, Bertus P, Szymoniak J (unpublished results)
16. Tsuji T, Nishida S (1987) In: Rappoport Z (ed) The chemistry of the cyclopropyl group. Wiley, New York, p 307
17. (a) Ruhlmann K (1971) Synthesis 236; (b) Rousseau G, Slougui N (1984) J Am Chem Soc 106:7283; (c) Nakamura E, Aoki S, Sekiya K, Oshino H, Kuwajima I (1987) J Am Chem Soc 109:8056; (d) Rieke RD, Sell MS, Xiong H (1995) J Am Chem Soc 117:5429; (e) Kasatkin A, Sato F (1995) Tetrahedron Lett 36:6079; (f) Thery N, Szymoniak J, Moïse C (2000) Eur J Org Chem 1483; (g) Taylor RE, Risatti CA, Engelhardt FC, Schmitt MJ (2003) Org Lett 5:1377
18. (a) Rousset CJ, Swanson DR, Lamaty F, Negishi E (1989) Tetrahedron Lett 30:5105; (b) Ito H, Taguchi T, Hanzawa Y (1992) Tetrahedron Lett 33:1295; (c) Ito H, Nakamura T, Taguchi T, Hanzawa Y (1995) Tetrahedron 51:4507
19. Davis JM, Whitby RJ, Jaxa-Chamiec A (1992) Tetrahedron Lett 33:5655
20. Negishi E, Huo S (2002) In: Marek I (ed) Titanium and zirconium in organic synthesis. Wiley-VCH, Weinheim, p 1
21. (a) Takahashi T, Kondakov DY, Suzuki N (1993) Tetrahedron Lett 34:6571; (b) Suzuki N, Kondakov DY, Kageyama M, Kotora M, Hara R, Takahashi T (1995) Tetrahedron 51:4519
22. Gandon V, Laroche C, Szymoniak J (2003) Tetrahedron Lett 44:4827
23. Eilbracht P (1997) In: de Meijere A (ed) Methods of organic chemistry, vol E17c. Houben-Weyl, Thieme, Stuttgart, p 2677 and references therein
24. (a) Jung ME, Gervay J (1991) J Am Chem Soc 113:224; (b) Sammes PG, Weller DJ (1995) Synthesis 1205
25. (a) Wipf P, Jahn H (1996) Tetrahedron 52:12853; (b) Lipshutz BH, Pfeiffer SS, Noson K, Tomioka T (2002) In: Marek I (ed) Titanium and zirconium in organic synthesis. Wiley-VCH, Weinheim, p 110
26. Tam W, Rettig MF (1976) J Organomet Chem 108:C1
27. (a) Harada S, Kowase N, Taguchi T, Hansawa Y (1997) Tetrahedron Lett 38:1957; (b) Harada S, Kowase N, Tabuchi N, Taguchi T, Dobashi Y, Dobashi A, Hansawa Y (1998) Tetrahedron 54:753
28. Gandon V, Szymoniak J (2002) Chem Commun 1308
29. Hammaecher C, Bertus P, Haudrechy A, Szymoniak J (unpublished results)
30. (a) Casey CP, Strotman NA, J Am Chem Soc 2004, 1699, 2004; (b) In this context, Eq. 1 in Scheme 21 doesn't represent a real transition structure but rather an ideogram to be compared with the reactions in Eqs. 2 and 3
31. Bertelo CA, Schwartz J (1976) J Am Chem Soc 98:262
32. Ito H, Kuroi H, Ding H, Taguchi T (1998) J Am Chem Soc 120:6623
33. Ito H, Taguchi T (1997) Tetrahedron Lett 38:5829
34. Sato A, Ito H, Taguchi T (2000) J Org Chem 65:918
35. Ito H, Sato A, Taguchi T (1999) Tetrahedron Lett 40:3217
36. Ito H, Sato A, Kusanagi T, Taguchi T (1999) Tetrahedron Lett 40:3397
37. Negishi E, Akiyoshi K, O'Connor B, Takagi K, Wu G (1989) J Am Chem Soc 111:3089
38. Takahashi T, Ishikawa M, Huo S (2002) J Am Chem Soc 124:388
39. Takahashi T, Xi Z, Kotora M, Xi C, Nakajima K (1996) Tetrahedron Lett 37:7521

40. Zweifel G, Miller JA (1984) Org React 32:375
41. (a) Dzhemilev UM, Ibragimov AG, Zolotarev AP, Muslukhov RR, Tolstikov GA (1989) Bull Acad Sci USSR, Div Chem Sci 38:1981; (b) Dzhemilev UM, Ibragimov AG, Azhgaliev MN, Zolotarev AP, Muslukhov RR (1994) Russ Chem Bull Int Ed 43:252
42. (a) Dzhemilev UM, Ibragimov AG, Zolotarev AP, Tolstikov GA (1989) Bull Acad Sci USSR Div Chem Sci 38:1324; (b) Dzhemilev UM, Ibragimov AG, Azhgaliev MN, Muslukhov RR (1994) Russ Chem Bull Int Ed 43:255
43. Dzhemilev UM, Ibragimov AG (1998) Russ Chem Bull 47:786
44. Dzhemilev UM, Ibragimov AG, Zolotarev AP, Muslukhov RR, Tolstikov GA (1990) Bull Acad Sci USSR Div Chem Sci 39:1071
45. Dzhemilev UM, Ibragimov AG, Khafizova LO, Ramazanov IR, Yalalova DF, Tolstikov GA (2001) J Organomet Chem 636:76
46. Dzhemilev UM, Ibragimov AG, Ramazanov IR, Luk'yanova MP, Sharipova AZ, Kalilov LM (2000) Russ Chem Bull 49:1086
47. Negishi E, Montchamp JL, Anastasia L, Elizarov A, Chouery D (1998) Tetrahedron Lett 39:2503
48. Wipf P, Kendall C, Stephenson CRJ (2001) J Am Chem Soc 123:5122
49. Wipf P, Kendall C (2002) Chem Eur J 8:1804
50. Wipf P, Kendall C, Stephenson CRJ (2003) J Am Chem Soc 125:761
51. Negishi E (1981) Pure Appl Chem 53:2333
52. Negishi E (1999) Chem Eur J 5:411

Stereoselective Synthesis of Dienyl Zirconocene Complexes

Nicka Chinkov · Ilan Marek (✉)

Department of Chemistry and Institute of Catalysis Science and Technology,
Technion – Israel Institute of Technology, Haifa 32000, Israel
chilanm@tx.technion.ac.il

1	Introduction	134
2	Dienylzirconocene Derivatives	135
2.1	Hydrozirconation Reaction of Enynes	136
2.2	Dienyl Zirconocenes via Zirconacyclopentadiene Derivatives	137
2.2.1	Preparation of Zirconacyclopentadiene Derivatives	137
2.2.2	Reactivity of Zirconacyclopentadiene Derivatives	141
2.3	Dienyl Zirconocenes via Carbenoid Derivatives	145
3	Limitations of the Previously Described Methodologies	148
4	Stereoselective Synthesis of Vinyl Zirconocene Derivatives	148
5	Stereoselective Preparation of Dienyl Zirconocene Derivatives Via a Tandem Allylic C–H Bond Activation–Elimination Sequence	152
	References	164

Abstract Several methods are known for the preparation of dienyl zirconocene derivatives such as the hydrozirconation of enynes, the stereospecific reaction of zirconacyclopentadiene derivatives with electrophiles, the reaction of organozirconocene derivatives with carbenoid reagents, and the reaction of vinyl zirconocenes with vinyl halides. All these described methodologies lead to the expected dienyl zirconocenes but the stereoselectivity is always structure dependent. On the other hand, dienyl zirconocenes can also be easily prepared, as unique geometrical isomers, from simple nonconjugated unsaturated enol ethers with (1-butene)ZrCp$_2$ complexes. This methodology is based on a tandem allylic C–H bond activation–elimination sequence and the mechanism has been mapped out by deuterium labeling experiments. The stereochemical outcome of this process was determined by addition of several electrophiles. When the organometallic derivative was vinylic as well as allylic, an unexpected reversal of the stereochemistry was found during the zirconium to copper transmetalation step.

Keywords Dienyl zirconocene · Zirconacyclopentadiene · Allylic activation · Isomerization · Elimination

1
Introduction

The preparation of metalated dienyl derivatives is still a challenging problem in organic synthesis as only a few examples are reported in the literature. Since dienyl metals could be very useful synthetic precursors, efficient preparative methodologies opening new routes to stereodefined and functionalized systems would have a large number of synthetic applications.

Apparently, the simplest approach would be the carbometalation reaction or, more specifically, the vinylmetalation of alkynes [1–8]. The addition of organometallic reagents to functionalized or nonfunctionalized, terminal or nonterminal alkynes, in which the resulting organometallic compound can react with electrophiles, is defined as the carbometalation reaction (Scheme 1). It has been widely explored and applied in the regio- and stereoselective preparation of numerous vinyl metal species.

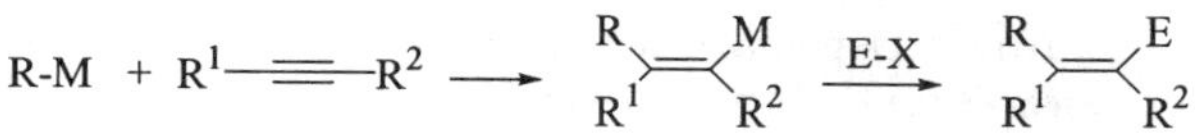

Scheme 1

Among the most prominent examples of this reaction are the carbocupration [5], the zirconium-catalyzed carboalumination [9–12], the nickel-catalyzed carbozincation [13–14], the allylmetalation of metalated alkynes [15], the allylzirconation [16], the allylgallation [17], the allylmanganation [18–19], and the alkyllithiation [20] reactions.

While the allylmetalation of alkynes is a well-described reaction, the addition of vinyl metal to a triple bond has remained comparatively unexplored [1–8, 21–25]. It stems from the fact that the metalated dienyl derivative 3, which results from the vinylmetalation of the alkyne 1, exhibits similar reactivity to that of 2 toward the alkyne 1. Therefore, 3 reacts also with the initial substrate 1 to give first the dimer, and after subsequent following additions, oligomers are formed (Scheme 2). Several alternatives were reported in the literature, and the most representative examples are described below.

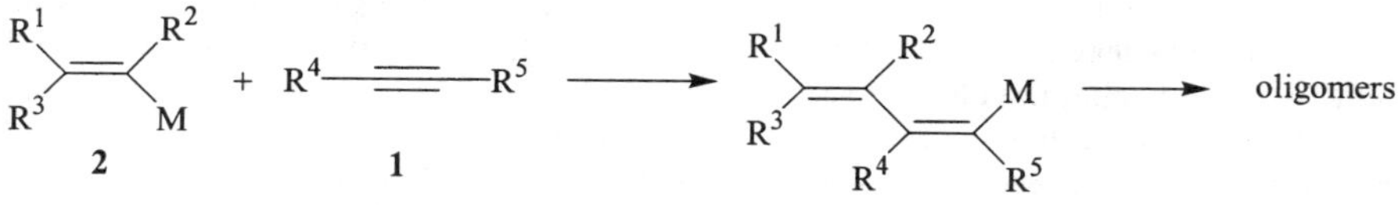

Scheme 2

Bromoboration of terminal alkynes 4 into the β-bromo-1-alkenyl-boronic esters 5 [26], followed by a palladium-catalyzed displacement of the β-halogen with organozinc reagents [27] is a known strategy for the preparation of metalated diene 6 (Scheme 3). An additional approach is the diboration of symmetrically disubstituted alkynes 7 with bis(pinacolato)diboron 8, followed

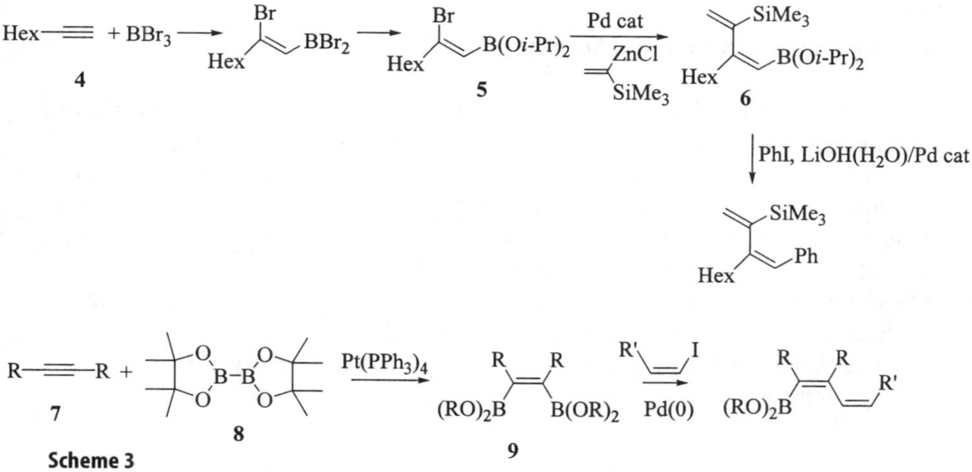

Scheme 3

by a selective coupling reaction of the resulting 1,2-bismetalated olefin **9** [28] (Scheme 3).

Bis(pinacolato)diboron **8** also reacts with 1-halo-1-lithioalkenes **10** (available from 1,1-dihaloalkenes or 1-haloalkenes by halogen–metal exchange or metalation reaction, respectively) to afford 1,1-bismetalated dienes **11**. These are readily converted into polysubstituted dienes through various transition metal-catalyzed carbon–carbon bond formations (Scheme 4) [29]. However, these seemingly attractive methods must not hide the problems of the stereochemistry and regiospecificity of the subsequent cross-coupling reactions.

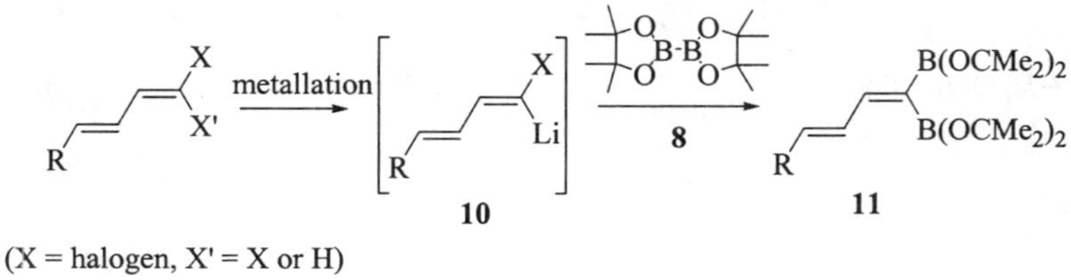

(X = halogen, X' = X or H)

Scheme 4

2
Dienylzirconocene Derivatives

As zirconocene derivatives constitute a powerful tool for various synthetic pathways, the preparation of zirconated dienes as synthetic precursors has triggered the development of a plethora of methodologies for their synthesis. In this chapter, we describe the existing approaches with a special emphasis on the problems of regio- and stereoselectivities as well as on their scope and limitations.

2.1
Hydrozirconation Reaction of Enynes

The addition of $Cp_2Zr(H)Cl$, known as the "Schwartz reagent" [30], to different alkenes and alkynes is known to be a facile process [31]. Therefore, the hydrozirconation of a variety of readily available enynes **12** is among the first methods developed for the stereoselective preparation of dienyl zirconium reagents **13**. This process is both completely chemo- and regioselective with a *syn* addition of the zirconium hydride across the alkyne [32] (Scheme 5). From the same intermediate, the Zr atom can be isomerized in its internal position such as in **15** via a zirconacyclopropene intermediate **14**. Moreover, the addition of trimethylstannyl chloride to **14** led to the stannylated dienyl zirconocene **16** [33] (Scheme 5).

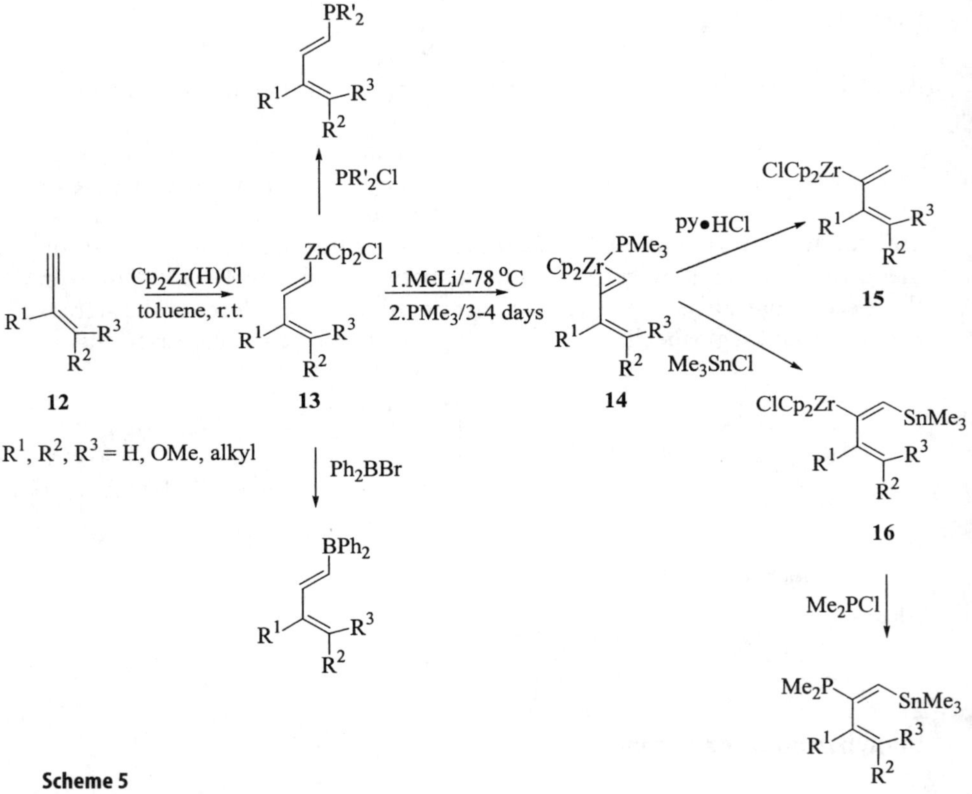

Scheme 5

2.2
Dienyl Zirconocenes via Zirconacyclopentadiene Derivatives

2.2.1
Preparation of Zirconacyclopentadiene Derivatives

Hydrozirconation of alkynes with the Schwartz reagent Cp$_2$Zr(H)Cl yields the chlorovinyl zirconocene **17**, which is easily converted to the methyl vinyl zirconocene **18** with either methyllithium in THF or methylmagnesium bromide in CH$_2$Cl$_2$. Compound **18** loses further methane at room temperature to form a zirconacyclopropene intermediate **19**, which couples with a second alkyne and forms the metallacyclopentadiene **20** (Scheme 6) [34].

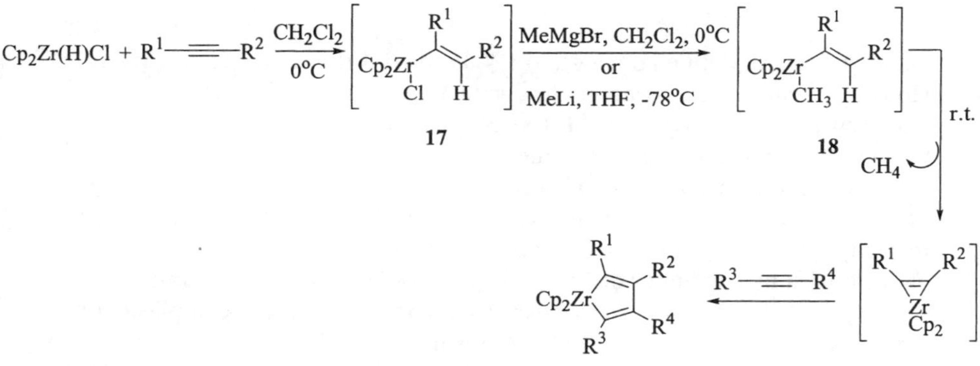

Scheme 6

Several asymmetrically substituted zirconacyclopentadienes were prepared by this methodology as unique regioisomers, as described in Fig. 1. If mixtures are initially formed, equilibration of the mixture to one isomer can be accomplished by heating the reaction to 80 °C. These metallacycles are cleaved with essentially complete retention of configuration by protonolysis or by iodinol-

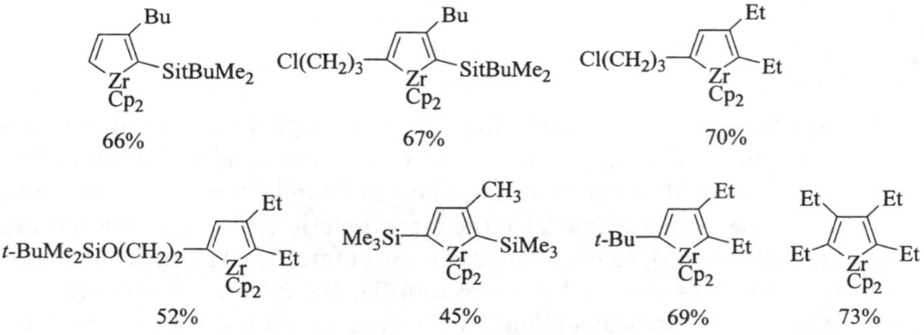

Fig. 1 Asymmetrically substituted zirconacyclopentadiene regioisomers

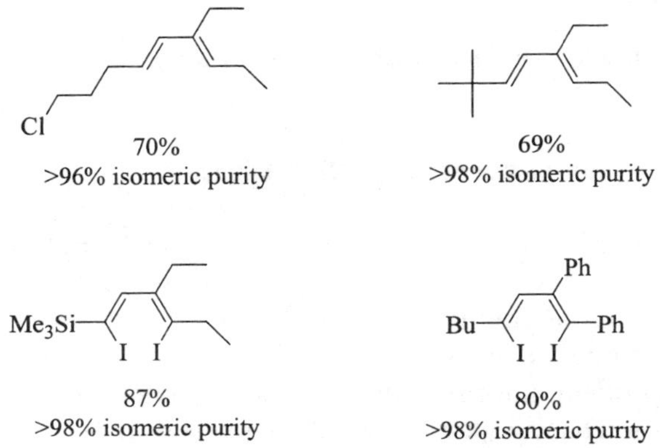

Fig. 2 Dienes and 1,4-diiodo-1,3-dienes from cleavage of metallacycles

ysis to give stereodefined dienes and 1,4-diiodo-1,3-dienes, respectively (Fig. 2). However, despite the easy access to diversely substituted zirconacyclopentadiene derivatives, no examples of the selective reaction of the carbon–zirconium bond with two different electrophiles were described in this study [35].

An alternative preparation of symmetrically substituted zirconacyclopentadiene 22 was also reported by the reaction of Cp_2ZrBu_2 21 (the so-called Negishi reagent) with two equivalents of alkyne. The Negishi reagent 21 can be easily prepared in situ by treatment of Cp_2ZrCl_2 with two equivalents of n-butyllithium [36–38]. It has been proposed to exist as a zirconocene–butene complex equivalent, Cp_2Zr^{IV} (as 21a) and Cp_2Zr^{II} (as 21b), and may be best viewed as resonance hybrids between the two species 21a and 21b, as depicted in Scheme 7. *In the following discussion, Cp_2ZrBu_2 21 will be used to describe the Negishi reagent.*

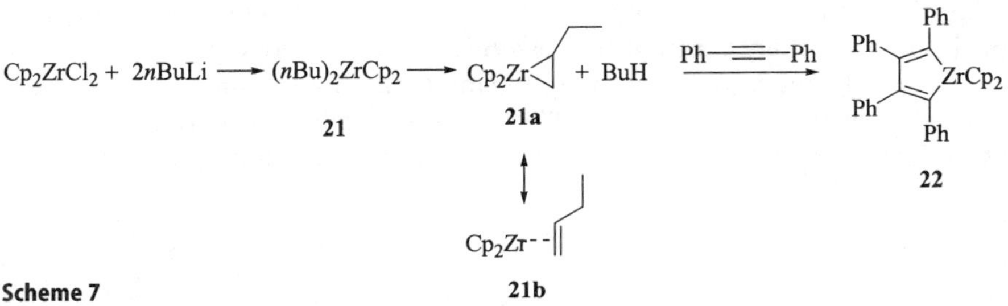

Scheme 7

It would be synthetically interesting to cross-couple two different unsymmetrical alkynes with complete regiocontrol. However, one limitation of this chemistry is that neither 1-hexyne nor 3-hexyne couples cleanly with a second terminal alkyne, which would give the largest steric difference between the ends of the alkyne [39]. To have very clean unsymmetrical zirconacyclopentadiene derivatives, the use of ethene is primordial. Indeed, an excess of ethylene leads first to a zirconacyclopentane 23, which can react successfully with two different alkynes to give 20 (Scheme 8).

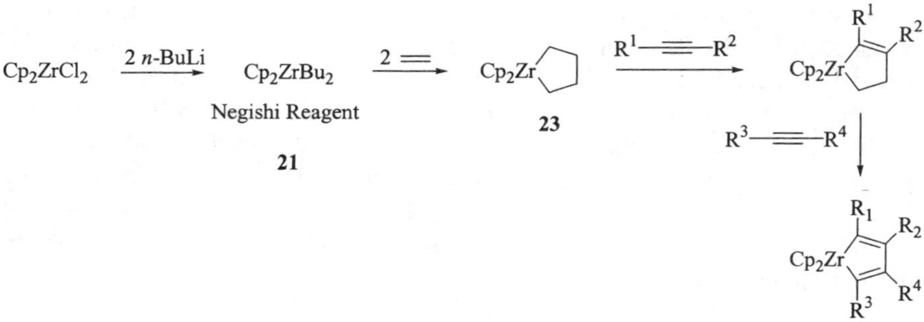

Stereoselective Synthesis of Dienyl Zirconocene Complexes

Scheme 8

An easier route to the same unsymmetric zirconacyclopentadiene **20** is the use of Cp$_2$ZrEt$_2$ as reagent (easily prepared from Cp$_2$ZrCl$_2$ with two equivalents of EtMgBr). This reagent (Scheme 9), equivalent to a zirconocene–ethene complex **24**, reacts similarly to the Negishi-type reagent, which was prepared in the

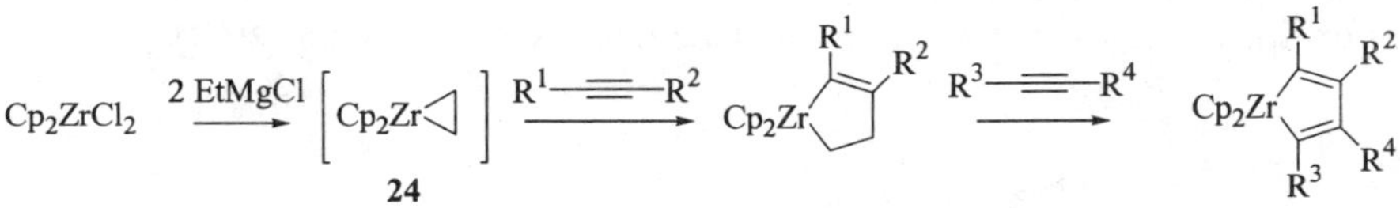

Scheme 9

presence of excess ethylene. Although this generalized scheme might suggest that the formation of zirconacyclopentadienes would be complicated with difficulties associated with "pair" selectivity and regioselectivity, many reactions are very selective: generally, alkyl substituents strongly prefer to be β to the Zr while aryl

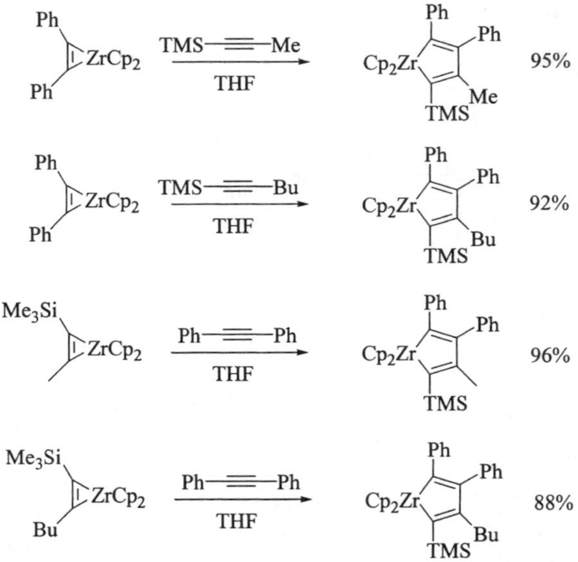

Scheme 10

and silyl groups strongly favor the α position. The former must be largely steric in origin, while the latter must be electronic. Representative examples are described in Scheme 10 [35, 40]. However, if two alkynyl compounds are similar in chemical properties, mixtures of zirconacyclopentadienes will be formed.

One can use the bicyclization of diynes as an alternative method for the regioselective formation of zirconacyclopentadienes. This strategy, initially using the combination of $Cp_2ZrCl_2/Mg/HgCl_2$ [41, 42] and later the Negishi reagent Cp_2ZrBu_2 [43] (Scheme 11), leads to the intramolecular carbocyclization of two alkynyl moieties.

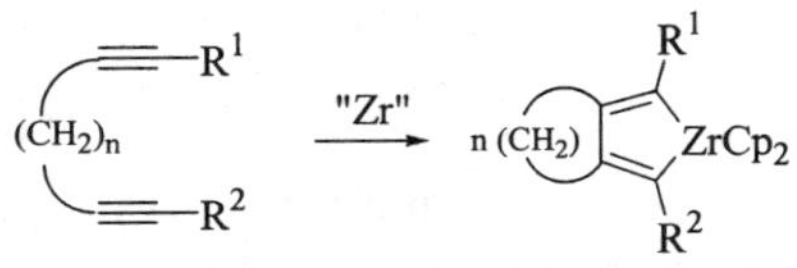

Scheme 11   n = 2–5   **25**

Both methods provide four-, five-, six-, and seven-membered-ring products, affording a single regioisomer of the zirconacycle **25**. The substrate can possess

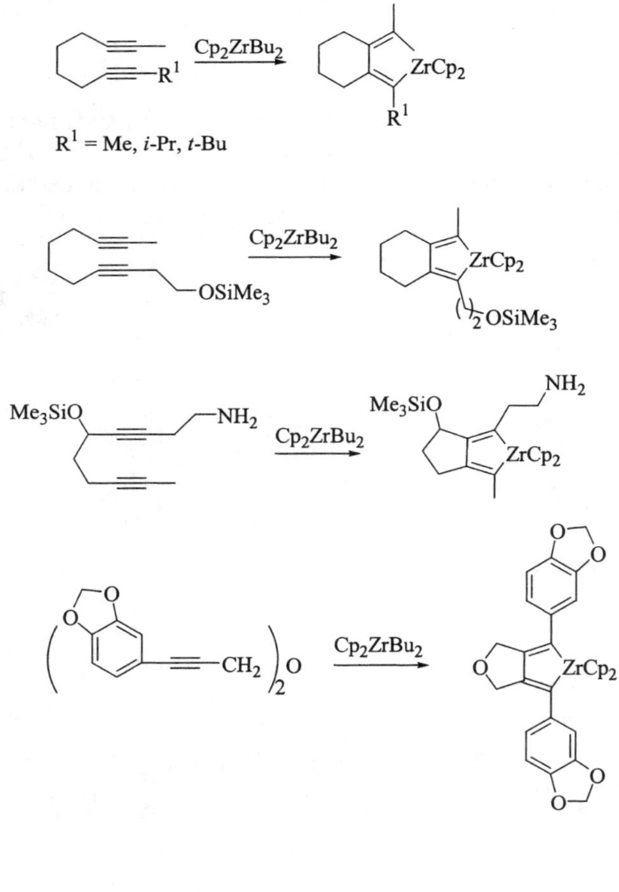

R^1 = Me, i-Pr, t-Bu

Scheme 12

not only R¹ and R² of either equivalent or different nature, but it was also shown that this methodology was compatible with different functional groups. The cyclization of several heteroatom-containing diynes was performed as described in Scheme 12. As for the limitations of the given method, terminal diynes were unable to be cyclized.

2.2.2
Reactivity of Zirconacyclopentadiene Derivatives

Once single zirconacyclopentadienes **20** are formed, only the regiospecific reaction of one carbon–zirconium bond with an electrophile could lead to the stereoselective preparation of metalated dienes **26** (Scheme 13).

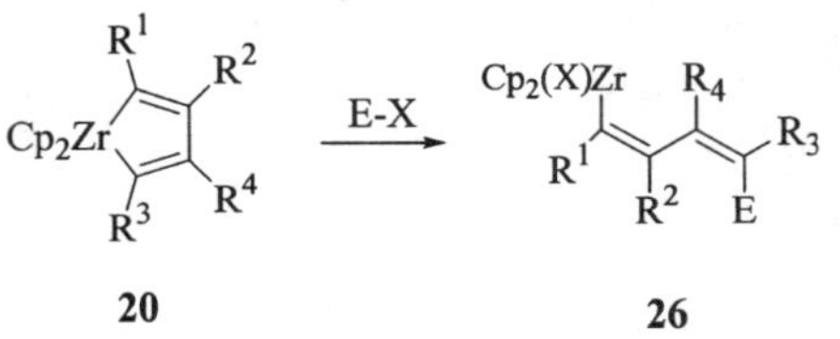

Scheme 13

The protonolysis of the equally substituted zirconacycle **27** with weak acids such as ethanol leads to the monoprotonated dienyl zirconocene compound **28**, which is subsequently converted into substituted diene **29** by a palladium cross-coupling reaction with aryl iodide (Scheme 14) [44]. Selective halogenation of such zirconacycle leads, as well, to the formation of zirconated diene **30** and then to the 1,4-dihalogenodiene **31** (Scheme 15) [45].

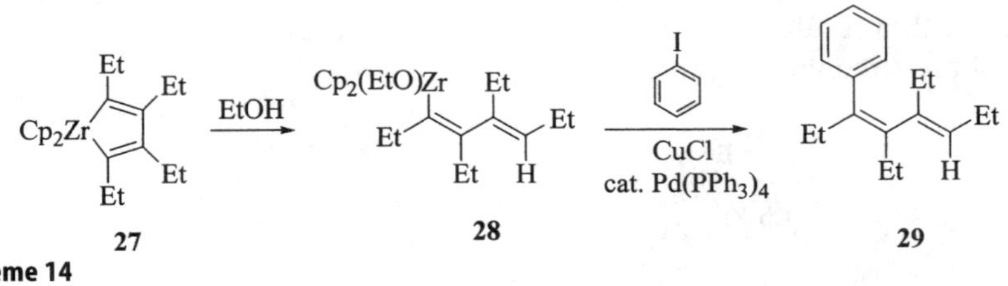

Scheme 14

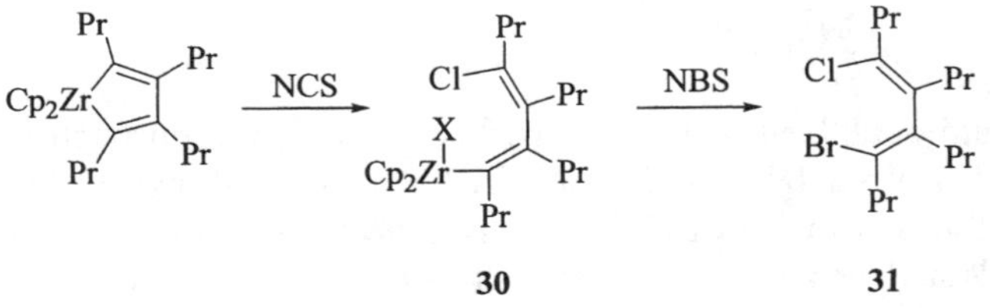

Scheme 15

On the contrary, once unsymmetrical zirconacyclopentadiene **20** (Scheme 13) is obtained regiospecifically, the formation of stereodefined metalated dienyl zirconocenes **26** requires a regiospecific reaction of only one carbon–zirconium bond (either R^1–C–Zr or R^3–C–Zr) of **20** with electrophiles (Scheme 13). Only a few examples were reported for this selective reaction and they are all related to substrates with two substitutents R^1 and R^3 of very different nature. Thus, chlorination of **32** with *N*-chlorosuccinimide (NCS) occurs at the methyl-substituted carbon attached to zirconium, while subsequent iodination of **33** takes place at the carbon-bearing phenyl substituent (Scheme 16) [45]. Similarly, propyl-substituted carbon is more reactive than the phenyl carbon–zirconium bond, as shown in the case of **34**. This excellent method, therefore, is substituent-dependent.

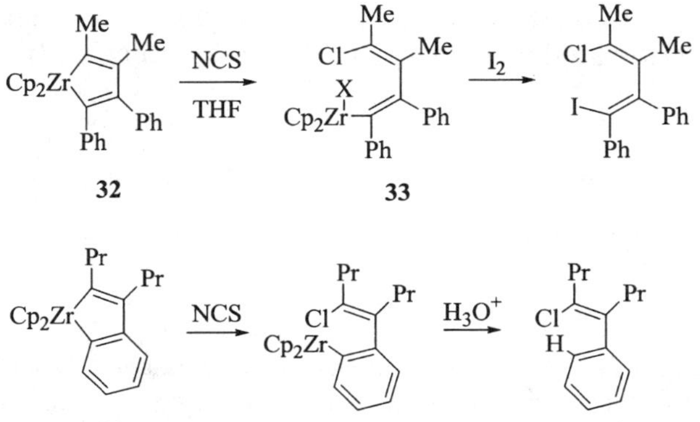

Scheme 16

On the other hand, transmetalation to copper and further reactions with electrophiles occur preferentially at the phenyl-bearing carbon of unsymmetrically substituted zirconacyclopentadiene derivative **35** (Scheme 17). The presence of lithium salts should be avoided to prevent the double allylation reaction of the zirconacyclopentadiene derivative [46].

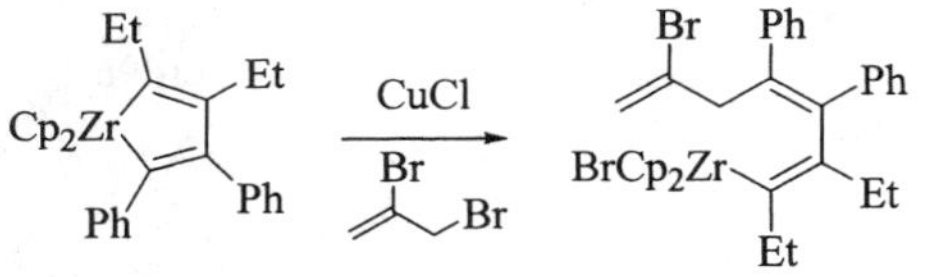

Scheme 17

Takahashi et al. have also developed an alternative approach to dienyl zirconocene by ethenylzirconation of vinyl ethers. As shown in Scheme 9, the reaction of alkynes with Cp$_2$ZrEt$_2$ **24** gives zirconacyclopentenes. Unsaturated compounds such as vinyl ethers can easily replace the ethylene moiety of the zirconacycle to afford potentially two regioisomers **37** and **38** [47–49] (Scheme 18). As **37** and **38** are in equilibrium, **38** undergoes a β-elimination reaction to give **39** as a unique isomer. This sequence may be considered as a vinylzirconation reaction of alkynes. Although substituted alkenyl ethers, such as **40** and **41**, did

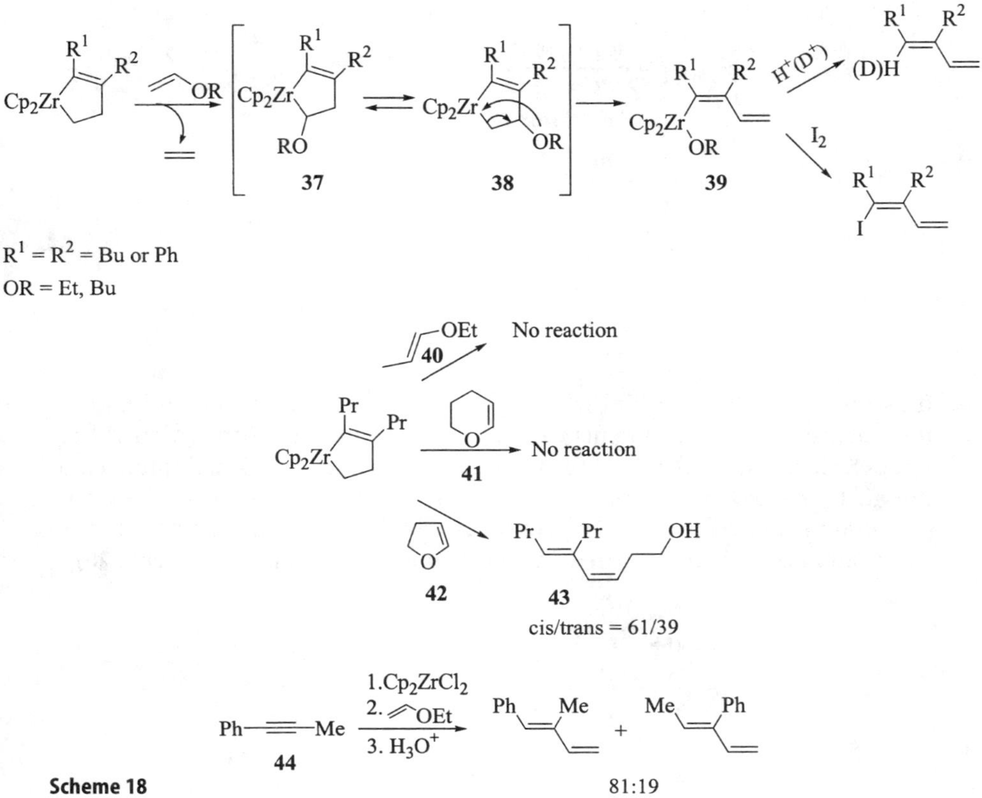

Stereoselective Synthesis of Dienyl Zirconocene Complexes 143

$R^1 = R^2 = Bu$ or Ph
OR = Et, Bu

Scheme 18

not give the corresponding products, 2,3-dihydrofuran **42** reacted smoothly to give **43** in 66% yield (*cis/trans*=61/39) after hydrolysis.

However, although with symmetrical internal alkynes such as 5-decyne and diphenylacetylene, vinylzirconation products were obtained with a good isomeric purity, unsymmetrical alkyne such as 1-propynyl benzene (R^1=Ph, R^2=Me) **44** gave a mixture of 81:19 regioisomers (Scheme 18). Obviously, if disubstituted alkynes with two different alkyl groups were used in this strategy ($R^1 \neq R^2$, alkyls) a mixture of isomers would be formed. The described methodology has also a second structural limitation as only a terminal double bond is formed in **39** (only **43** was successfully prepared, but as two isomers).

Barluenga et al. have demonstrated that the reaction of organolithium compounds **45** with zirconocene methyl chloride in THF, followed by addition of different vinyl bromides and further heating to +65 °C, led to dienes **46** and **47** in different ratios (Scheme 19) [50]. The latter was demonstrated to be dependent on the structure of the starting organolithium compound and of the vinyl bromide used. Thus, with the use of nonsubstituted vinyl bromide **48**, a mixture of regioisomeric dienes **49** and **50** was obtained, the "branched" one being the major isomer (Scheme 20). A reverse ratio was obtained for the *trans-β*-bro-

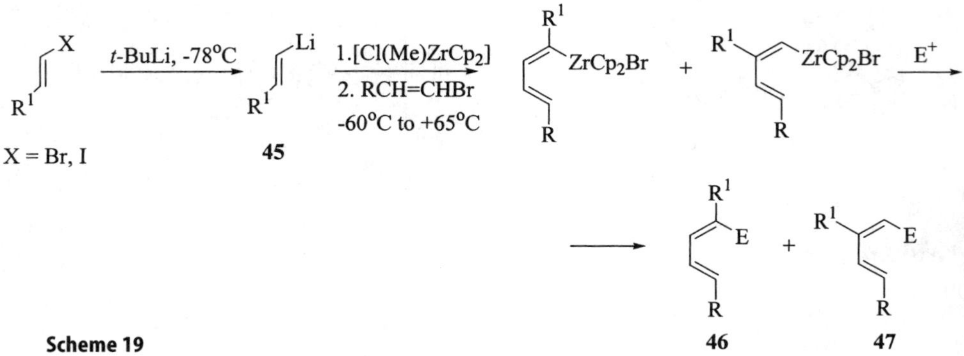

Scheme 19

mostyrene 51, affording "linear" dienyl zirconocene 52 as the major regioisomer. Interestingly, experiments carried out with alkyl-substituted vinyl bromides 54 led, independently of the structure of the starting organolithium compound, to the exclusive formation of the corresponding "branched" dienes 55. The second isomeric diene 57 was uniquely obtained while styrenyl lithium 56 was used in combination with a nonsubstituted vinyl bromide (Scheme 20).

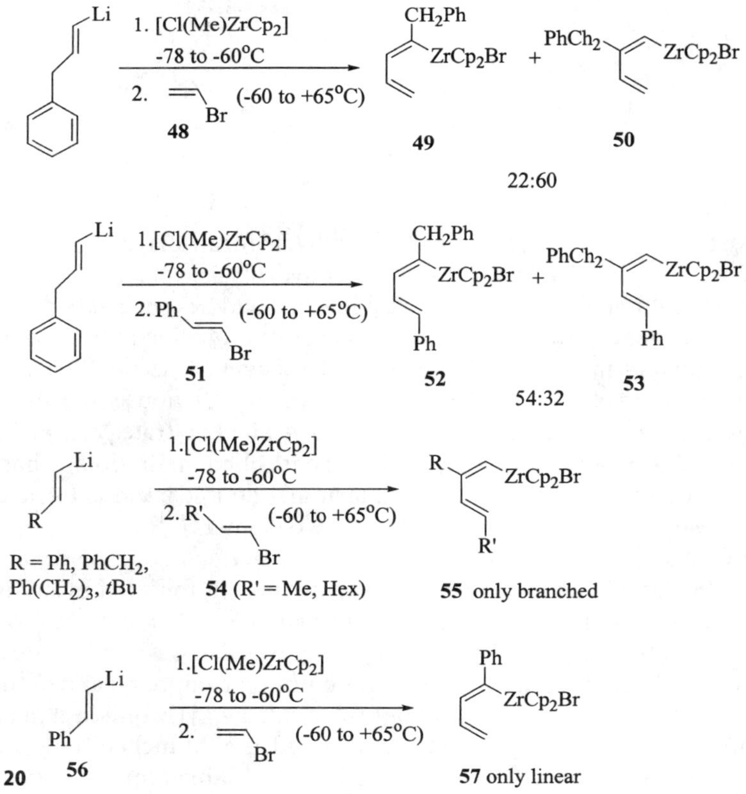

Scheme 20

Butyl vinyl ether **58**, coupled with zirconocene complex, led as well to the single "linear" regioisomer of the diene **59** (Scheme 21).

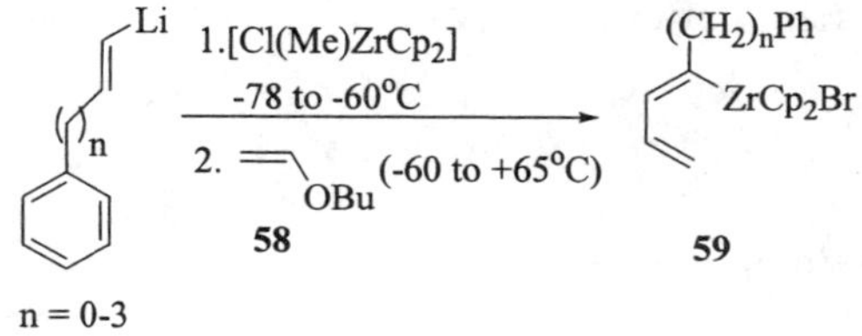

Scheme 21 n = 0-3

The above-mentioned results were presumably rationalized by the combination of electronic and steric effects. The reaction conditions were found to be particularly important, since maintaining the reaction temperature at 25 °C led to the formation of a mixture of products: cyclobutenyl derivative **61** was formed along with the "branched" dienyl moiety **60** (in contrast to reaction at +65 °C, which afforded dienyl products uniquely, Scheme 22).

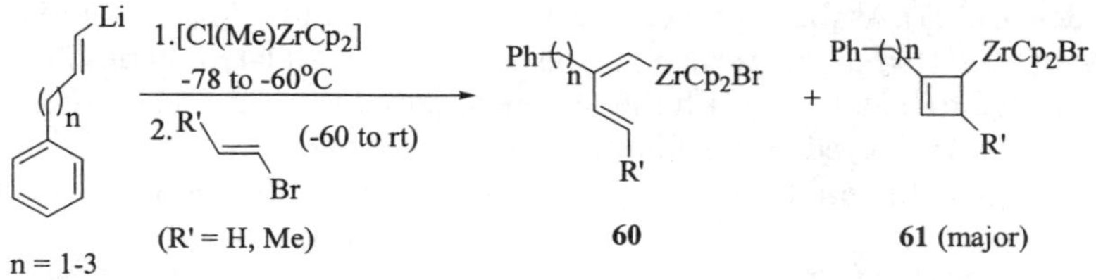

Scheme 22

2.3
Dienyl Zirconocenes via Carbenoid Derivatives

R.J. Whitby has recently described a new and elegant synthesis of nonterminal metalated dienes, based on 1-halo-1-lithioalkene insertion into acyclic zirconocene chlorides. Alkenyl carbenoids **63** are generated by halogen/lithium exchange at low temperature or by deprotonation of alkenyl halides (Scheme 23). The insertion of zirconocene chloride **62** into a carbenoid may be carried out in situ, thereby reducing the need for very low temperatures. Initial attack of the carbenoids **63** on the 16-electron zirconium atom of **62** to form 18-electron metalate complex **64** may be followed by a 1,2-rearrangement with loss of halogen to give a new organozirconocene **65**. The overall process is accompanied by a clean inversion of configuration at the carbenoid carbon [51].

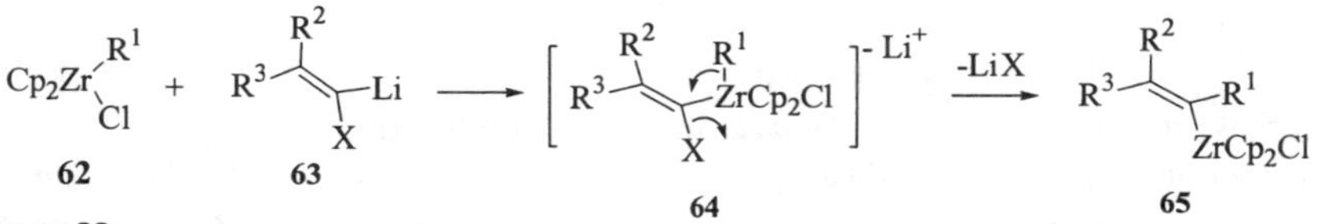

Scheme 23

When the above-mentioned process is applied to 1-lithio-1-chloroethene **66** [52] and vinyl zirconocene **67**, arising from hydrozirconation of terminal alkynes, a stereospecific formation of 2-zirconated dienes **68** is obtained, affording terminal dienes after protonation [53] (Scheme 24).

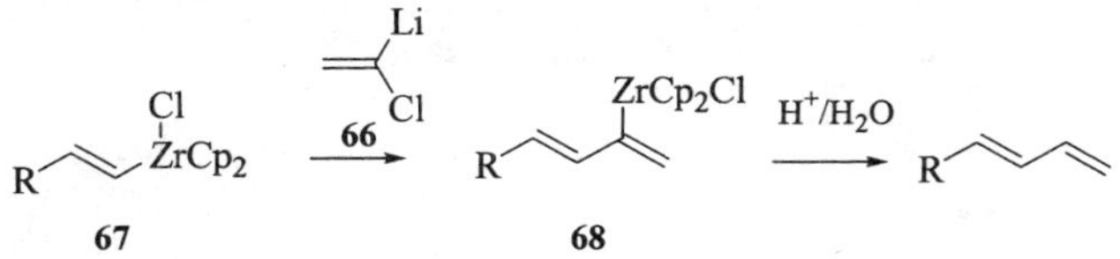

Scheme 24

The stereospecific insertion of 2-monosubstituted alkenyl carbenoids was successfully employed in the preparation of 1-alkyl-1-zircono-dienes. The Z and E carbenoids of 1-chloro-1-lithio-1,3-butadiene (**69** and **70**, respectively) are generated in situ from E- and Z-1,4-dichloro-2-butene [53] (Scheme 25). Inversion of configuration at the carbenoid carbon during the 1,2-metalate rearrangement stereospecifically yields terminal dienyl zirconocenes **71** and **72** [54] (Scheme 25). As the carbenoid-derived double bond is formed in ~9:1=Z:E for **69** and >20:1=E:Z isomeric mixtures for **70**, the metalated dienes **71** and **72** are expected to be formed with the same isomeric ratio. Carbon–carbon bond formation was achieved by palladium-catalyzed cross-coupling with allyl or vinyl halides to give the functionalized products with >95:5 stereopurity [55–57].

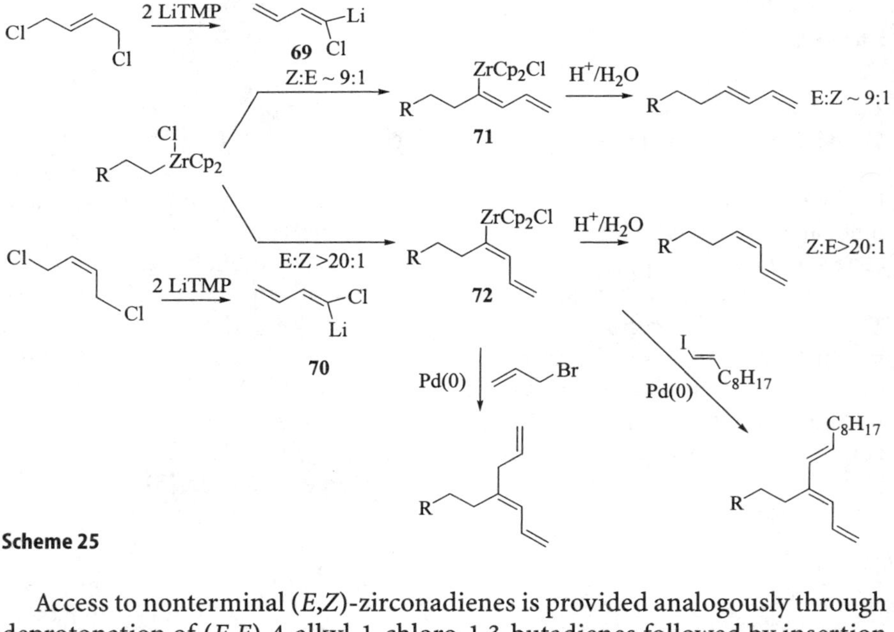

Scheme 25

Access to nonterminal (E,Z)-zirconadienes is provided analogously through deprotonation of (E,E)-4-alkyl-1-chloro-1,3-butadienes followed by insertion of the resultant carbenoid **73** into alkylzirconocene chlorides [53] (Scheme 26).

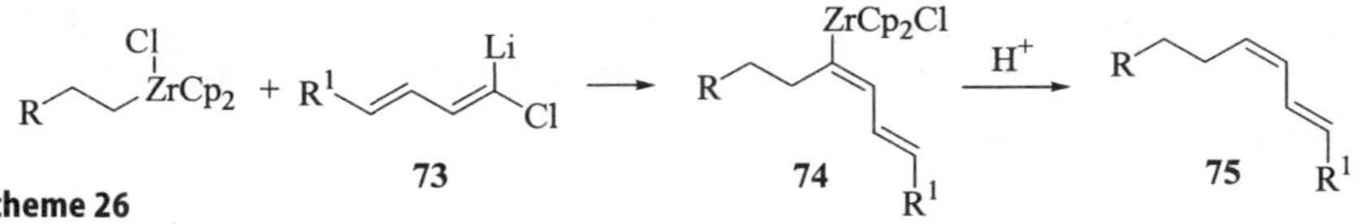

Scheme 26

Further hydrolysis of **74** gives the nonterminal (*E,Z*)-dienes **75** with high yields and stereoselectivities.

Unlike the insertion of 2-monosubstituted alkenyl carbenoids (**69**, **70**, and **73**), the reaction of 2,2-disubstituted alkenyl carbenoids with alkenyl zirconocene chlorides afforded the expected products as a mixture of stereoisomers. Thus, when **77**, derived from the deprotonation of the stereodefined *E*-1-chloro-2-methyl-1-octene **76**, was reacted with *E*-1-hexenylzirconocene chloride **78** at low temperature, a partial inversion of configuration at the alkenyl carbenoid center occurred before or during the rearrangement to afford the expected metalated diene **79** with an *E:Z* isomeric ratio of 58:42 after hydrolysis (see **80**, Scheme 27) [53]. The poor stereocontrol was attributed to the "metal-assisted ionization" [58–60], which promotes the interconversion of the *E*- to the *Z*-alkenyl carbenoids **77**. The latter occurs at a rate comparable with that of the insertion into organozirconocene chloride, and hence this is responsible for the loss of stereochemistry.

The methodology described above leads to the stereospecific formation of terminal 2-zirconadienes (**68**, Scheme 24), terminal 1-alkyl-1-zirconadienes (**71** and **72**, Scheme 25), and to the 1,4-disubstituted dienylzirconocenes **74** (Scheme 26) of fixed geometry, while the stereochemistry of the 1,1-disubstituted 2-zirconadiene **79** (Scheme 27) is not controlled.

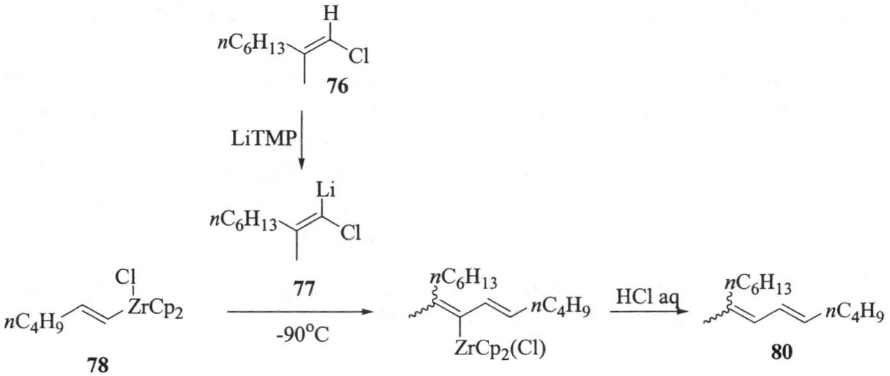

Scheme 27

A different approach for the in situ preparation of 2-zirconated 1,3-dienes **68** was reported by Szymoniak and Bertus et al., by treatment of 2-silyloxy-1,3-dienes **81** with dialkylzirconocene **21** (Scheme 28) [61]. The zirconocene induced the C–O bond cleavage of **81** and led to the dienyl zirconium compound **68**, which was used in dienylation reactions. Various cross-coupling reactions of **68** have been carried out with complete regioselectivity at C-2 and with moderate to good yields (Scheme 28).

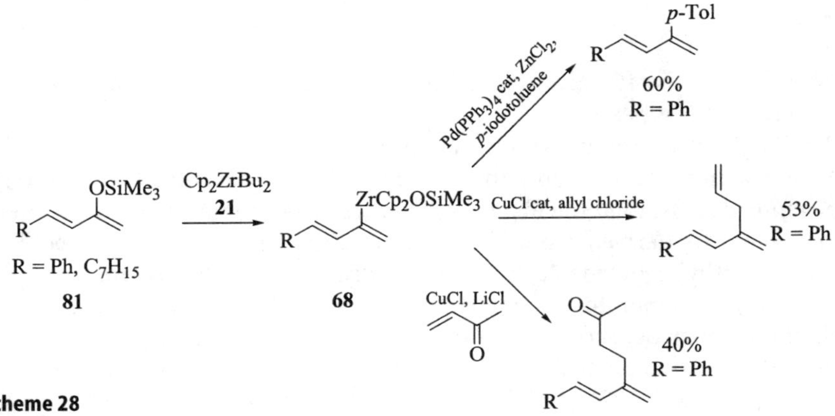

Scheme 28

3
Limitations of the Previously Described Methodologies

The impressive results outlined above are perhaps the most compelling examples of the power of zirconocene compounds in organic synthesis but in a general sense, are also illustrative for the serious limitations of the methodologies for the stereoselective preparation of dienyl zirconocene derivatives. These limitations can be summarized as follows:

1. R^1 and R^2 as well as R^3 and R^4 in **20** have to be of different electronic nature (Scheme 13).
2. Only unsubstituted terminal double bonds can be prepared by using the strategy described in Scheme 18 (only stereoselective preparation of **39**).
3. Problems of stereoselectivity arise when unstable carbenoids have to be used (except in the methodology leading to **74**, Scheme 26).
4. Stereoselectivity of the reaction could not be addressed in the preparation of 1-methylene-2-propenylzirconium (Scheme 28).

Therefore, new efficient methodologies are still required for the regio- and stereoselective preparation of dienylzirconocene derivatives.

4
Stereoselective Synthesis of Vinyl Zirconocene Derivatives

It has been recently reported that the treatment of heterosubstituted olefins such as **82a–g** [XR=OMe (**82a**), OSiMe$_2$Bu-*t* (**82b**), SPh (**82c**), SPr (**82d**), S(O)Tol (**82e**), SO$_2$Me (**82f**), SO$_2$Ar (**82g**)] with the Negishi reagent **21** led to the stereoselective synthesis of vinyl zirconocene derivatives 83 as described in Scheme 29. The formation of discrete organometallic derivatives was first checked by deuterolysis and iodinolysis, and then by creation of a carbon–carbon bond after the transmetalation reaction of the vinyl zirconocene 83 into vinyl copper derivatives.

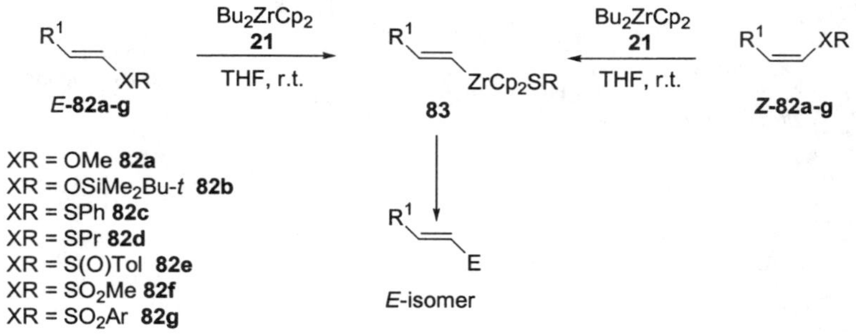

XR = OMe **82a**
XR = OSiMe$_2$Bu-*t* **82b**
XR = SPh **82c**
XR = SPr **82d**
XR = S(O)Tol **82e**
XR = SO$_2$Me **82f**
XR = SO$_2$Ar **82g**

Scheme 29

When the same reaction was performed on the Z isomer **82a–g** of the heterosubstituted alkenes, only the E-adduct **83** was obtained (determined after reaction with an electrophile; reaction of vinyl zirconocene with electrophiles occurs with retention of configuration). So, whatever the stereochemistry of the initial heterosubstituted alkenes, the reaction is >99% stereoselective but not stereospecific, producing only the E-vinyl zirconium in good overall yields. No stereoisomerization of Z-**82a–g** into E-**82a–g** was observed in the process, which indicates that zirconocene **21** is not a catalyst for the isomerization of the starting alkene. If we consider simple ligand exchange, different geometrical isomers of the corresponding vinyl zirconium derivatives **83** should be obtained when starting from the E-**82a–g** or Z-**82a–g** of vinyl-XR. Therefore, more complicated intermediates are most probably involved during the complexation between the olefins **82a–g** and the zirconocene **21**, since we have a complete isomerization reaction.

The first hypothesis was that the initial step proceeds via a dipolar zirconate species [62] represented by **84**, followed by an isomerization reaction leading to the *trans*-zirconacyclopropane **85** [63]. Then, after a β-elimination step, the corresponding E-vinyl zirconium should be obtained. Although this mechanistic interpretation was attractive, a stereochemical problem remains for the elimination reaction, since a dihedral angle of 180° (*anti* elimination) or 0° (*syn* elimination) is usually required for an elimination reaction. In this case, an angle of 120° is expected in the zirconacyclopropane **85** (Scheme 30). Moreover, this mechanistic interpretation does not explain the isomerization reaction of vinyl sulfone, since no dipolar zirconate intermediate can be present in this case. This isomerization was therefore explained by a carbometalative ring expansion between **21** and **82a–g**, leading to the corresponding five-membered ring zirconacycle **86** which may produce the three-membered ring zirconacyclopropane **85**, since facile equilibration among three- and five-membered ring zirconacycles has already been discussed for skeletal rearrangements [64].

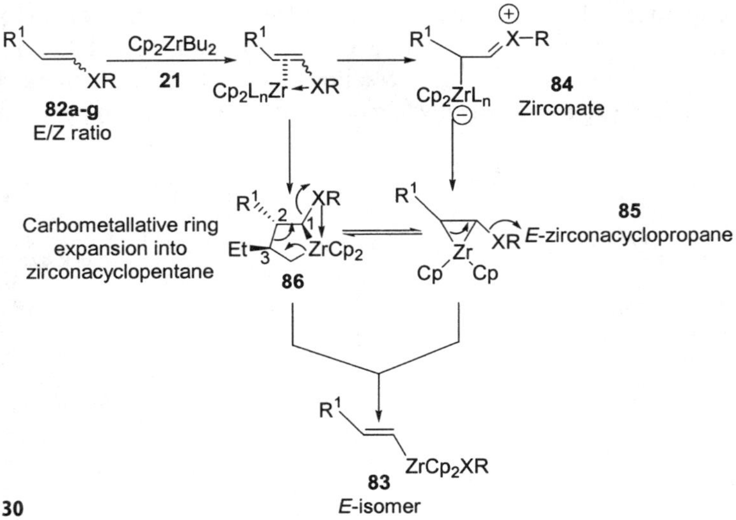

Scheme 30

Furthermore, if we consider the carbometalative ring expansion to produce the corresponding five-membered ring zirconacycle **86**, the carbon–heteroatom bond of the sp³ metalated center C₁ should isomerize to produce the most stable intermediate. Such isomerization could be due to an interaction between the heteroatom moiety XR and the zirconium atom [65], which would produce a weakness of the C₁–Zr bond and would facilitate the isomerization. Thus, whatever the stereochemistry of the starting material, a conformation is always possible in which C₁–SR is antiperiplanar to C₂–C₃ in **86** with a trans relationship between R′ and the ZrCp₂ fragment. The elimination reaction, or decarbozirconation, occurs in a concerted way to give the *E*-vinyl zirconium **83**. Unfortunately, neither the zirconacyclopentane nor the zirconacyclopropane have been trapped as intermediates.

On the other hand, Cp₂ZrEt₂ **24** (Scheme 9) has a different behavior, in most cases, from the butene moiety of the zirconocene–butene complex **21** [66]. Indeed, the ethylene ligand reacts with various unsaturated compounds and, as it is usually incorporated in the reaction products, we were interested to see if it could be also incorporated into heterosubstituted alkenes **82a–g**. Cp₂ZrEt₂ was easily prepared by treatment of two equivalents of EtMgBr (and also by two equivalents of EtLi to check that there is no salt effect) with Cp₂ZrCl₂ to furnish **24**, which was then treated with **82d** and **82f** (XR=SPr and SO₂Me, respectively) at room temperature. The addition product **87**, rapidly observed by gas chromatographic analysis of aliquots after hydrolysis, is followed by the elimination reaction to give the vinyl zirconium species **83** and ethylene. Both of the intermediates **88** and the product **89** were trapped by hydrolysis (Scheme 31). Although the formation of the vinyl zirconium **83** is slower in this particular case (only 2–3 h were necessary for the formation of **83** from **82d,f** and **21**), we can clearly see that the addition product **87** undergoes a subsequent

Stereoselective Synthesis of Dienyl Zirconocene Complexes

elimination reaction to give the expected product 83. Thus, from these mechanistic studies, it is believed that the unique formation of the *E* isomer 83 results from a carbometalative ring expansion into zirconacyclopentane followed by an elimination reaction.

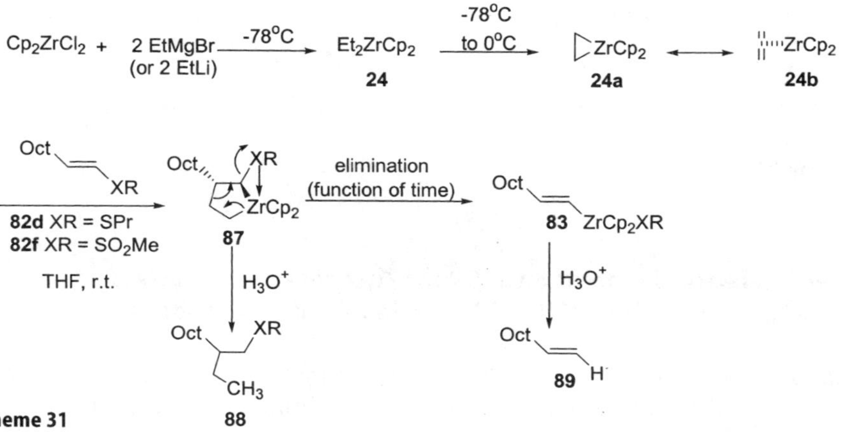

Scheme 31

Coming back to the initial problem, namely the stereoselective preparation of dienyl zirconocene derivatives, it was initially thought to apply the strategy described above to the preparation of dienyl systems (transformation of heterosubstituted conjugated dienes 90 into dienyl zirconocenes 91, as described in Scheme 32). A few examples will be shown later in the chapter.

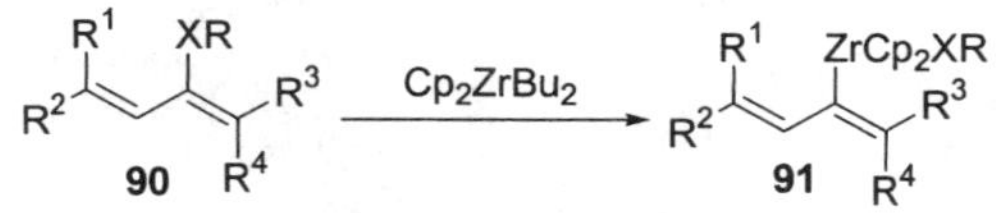

Scheme 32

On the other hand, the transition metal-catalyzed isomerization of terminal olefins into internal olefins has been extensively studied, and in general a mixture of 1-alkenes and *E*- and *Z*-2-alkenes, reflecting the thermodynamic equilibrium, is obtained [67]. Some low-valent titanocene derivatives are highly effective and stereoselective in favor of the *E*-2 isomer [68]. When non-conjugated dienes such as 92, containing one or two substituted vinyl groups, are treated with the zirconocene 21, a regioisomerization of the less-substituted

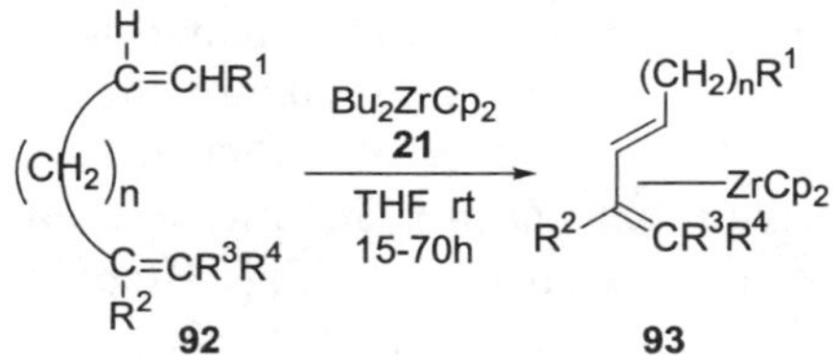

Scheme 33

double bond can occur and lead to the formation of the conjugated diene–zirconocene complexes **93** (Scheme 33) [69].

So, the combination of the isomerization process described in Scheme 33 with the elimination reaction described in Scheme 29 would lead to an efficient preparation of stereodefined metalated zirconocene complexes **95** from simple unsaturated enol ether **94** (Scheme 34).

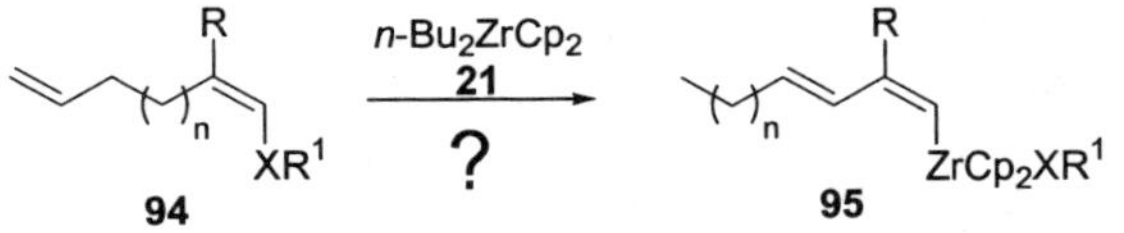
R
n-Bu2ZrCp2
21
?
XR1
R
ZrCp2XR1
n
n

Scheme 34 **94** **95**

5
Stereoselective Preparation of Dienyl Zirconocene Derivatives Via a Tandem Allylic C–H Bond Activation–Elimination Sequence

All the starting materials were very easily prepared in a single-pot operation by treatment of the alkoxy-allene **96** [70] with lithium organocuprate either in Et_2O (for the formation of the *Z*-vinyl copper intermediate *Z*-**97**) or in THF (for the formation of the *E*-vinyl copper intermediate *E*-**97**) and trapping the resulting alkenyl copper *E*- and *Z*-**97** with different unsaturated alkyl halides to give **98a–j** (Scheme 35) [71].

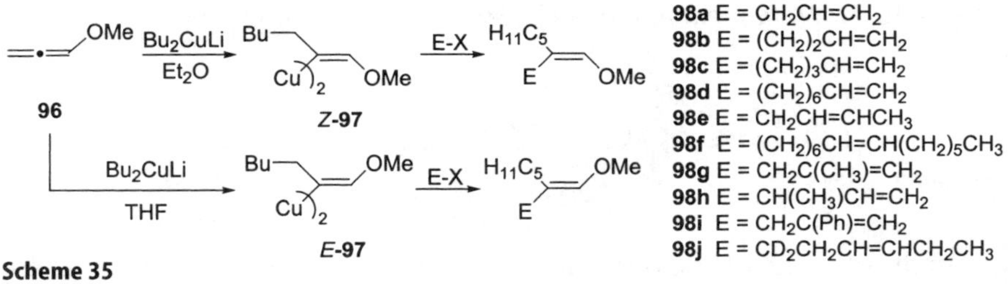
OMe
Bu2CuLi
Et2O
96
Bu
Cu
2
OMe
Z-97
E-X
H11C5
E
OMe
Bu2CuLi
THF
Bu
OMe
Cu
2
E-97
E-X
H11C5
OMe
E
98a E = CH2CH=CH2
98b E = (CH2)2CH=CH2
98c E = (CH2)3CH=CH2
98d E = (CH2)6CH=CH2
98e E = CH2CH=CHCH3
98f E = (CH2)6CH=CH(CH2)5CH3
98g E = CH2C(CH3)=CH2
98h E = CH(CH3)CH=CH2
98i E = CH2C(Ph)=CH2
98j E = CD2CH2CH=CHCH2CH3

Scheme 35

Originally, **98a** was treated with (1-butene)$ZrCp_2$ **21** in THF at room temperature and the evolution of the reaction was followed by gas chromatography of hydrolyzed aliquots and by ^1H NMR of the reaction after hydrolysis. This examination indicated that the starting material was consumed within 12 h with concomitant formation of several products including the expected diene **99** as the major product (with a maximum of 60% yield after hydrolysis, Scheme 36 path A). On the other hand, when **98a** was treated with **21** either in THF at +50 °C for 15 min or in Et_2O at +35 °C for 30 min, the corresponding diene **99** was constantly obtained with a yield of 80% (Scheme 36, path B).

The ^1H NMR spectrum of **99Zr** in C_6D_6/Et_2O solution showed two singlet peaks at 6.59 and 5.8 ppm assigned to the Zr–CH(sp^2) moiety and to the Cp protons, respectively, and a multiplet at 6.3 and 5.8 ppm for the hydrogens of

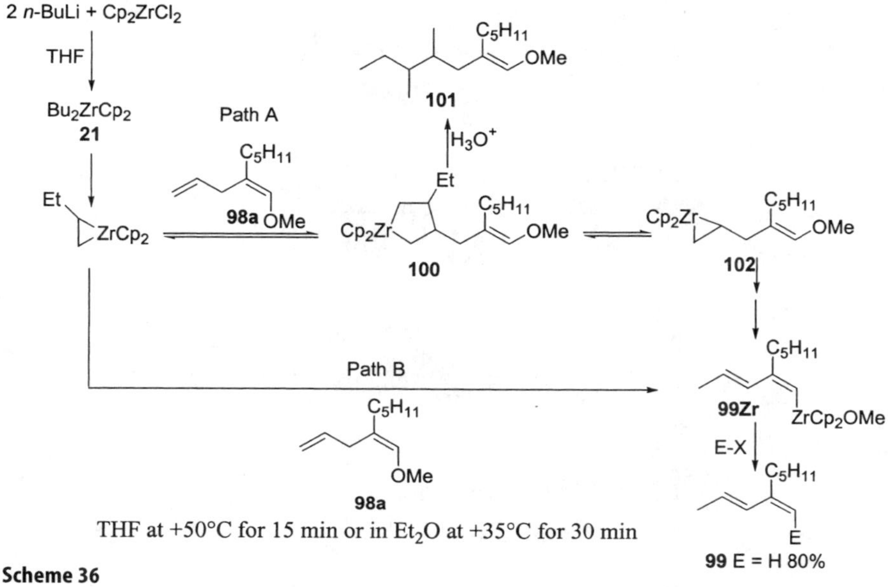

Stereoselective Synthesis of Dienyl Zirconocene Complexes

Scheme 36

THF at +50 °C for 15 min or in Et$_2$O at +35 °C for 30 min

the nonmetalated double bond. Its ^{13}C NMR spectrum revealed two singlets at 173.4 and 110 ppm which were assigned to the Zr–C(sp^2) moiety and Cp, respectively, and two more sp^2 carbons at 138.2 and 120.3 ppm.

Interestingly, only in the experiments done at room temperature (Scheme 36, path A) the addition product **101** and dimers of **100** were detected, both attributed to the putative intermediate **100** (Scheme 36). Then, this zirconacycle intermediate disappears in favor of the dienyl zirconocene complex, likely via the formation of (alkene)zirconocene **102** through a skeletal rearrangement [72] and then isomerization to lead to **99Zr**. These results support an associative mechanism involving the intermediary formation of monocyclic zirconacyclopentane. When the reaction is performed at higher temperature (reflux of solvent), the equilibrium between **100** and **102** is very rapidly displaced in favor of **102**, which can cleanly undergo the isomerization process. The latter experimental condition will be further generally used for the preparation of dienyl zirconocenes [73] (see scope of the reaction in Scheme 37).

The presence of a discrete organometallic species as well as the stereochemistry of the metalated diene were first checked by iodinolysis and bromolysis, but the corresponding iodo- and bromodienes were found to be unstable and rapidly isomerized to a mixture of E and Z isomers. Thus, in order to have a better picture of the stereochemical outcome of the process, the crude reaction mixture was treated with N-chlorosuccinimide (Scheme 37), and the corresponding chlorodiene **103** was isolated in 60% yield with an isomeric ratio >98:2. As alternative solution, the stereochemistry of the reaction was also determined by addition of allyl chloride, in the presence of a catalytic amount of copper salt, to the dienyl zirconocene derivative [74]. Skipped triene **104** was

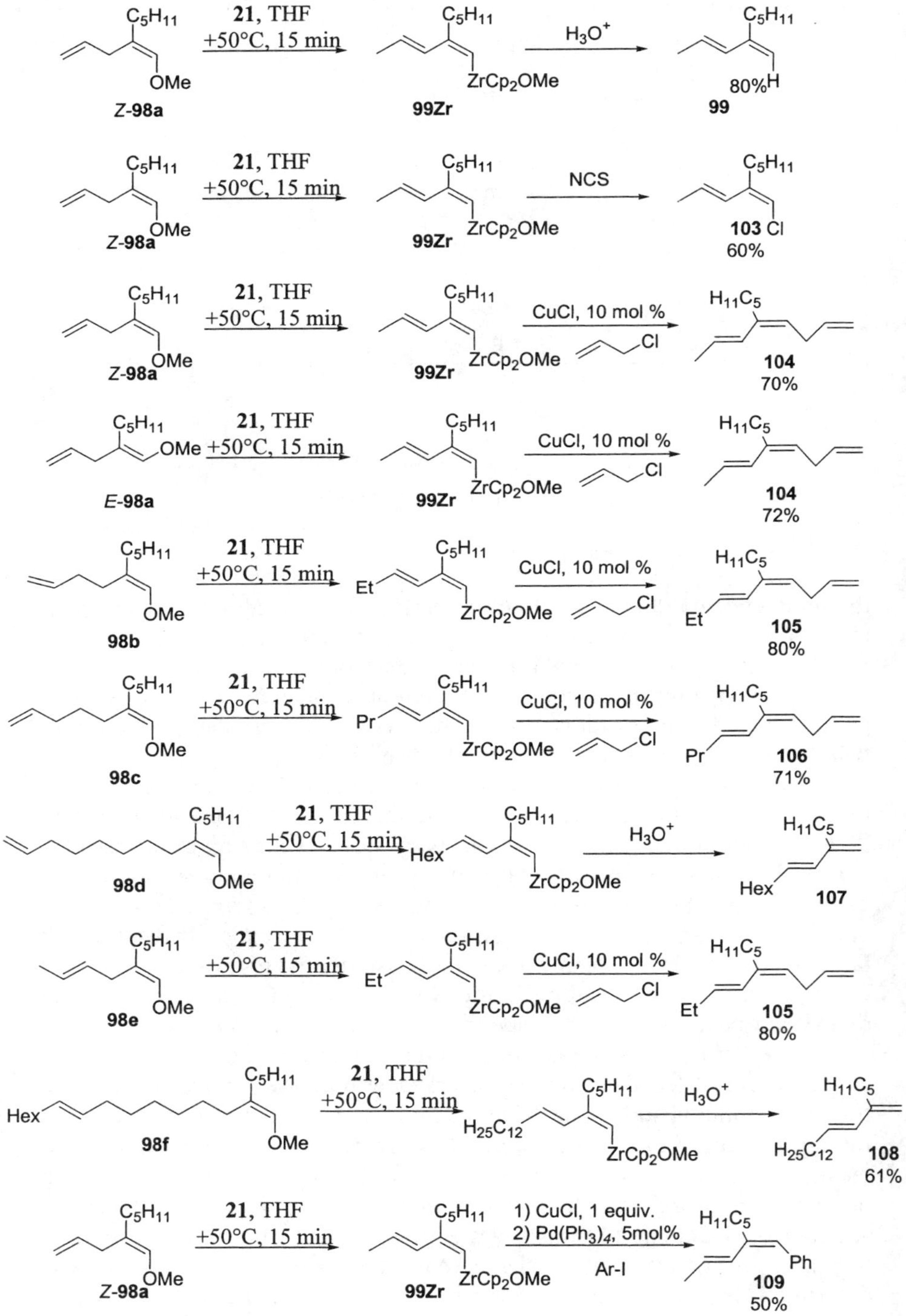

Scheme 37

obtained in good overall yield with an isomeric ratio greater than 98:2 in all cases. The (*E,Z*) stereochemistry of the 5-pentyl-octa-1-4*Z*,6*E*-triene **104** was deduced on the basis of differential nuclear Overhauser effect (NOE) spectra. When the same reaction was performed on the opposite isomer of the enol ether, namely the *E* isomer ***E*-98a** (prepared as described in Scheme 35) [71], the same (*E,Z*)-dienyl metal **99Zr** was obtained, as determined by the stereochemistry of the resulting product after reaction with allyl chloride (Scheme 37). So, whatever the stereochemistry of the starting enol ether, a unique isomer of the dienyl zirconocene is obtained at the end of the process.

Further on, a mixture of (*E,Z*) isomers will be used as starting ω-ene-enol ether **98a–h**. The formation of dienyl zirconocenes is not limited to those dienes with a one-carbon tether (Scheme 37). Compounds **98b–d** (with two-, three-, and six-carbon tethers, respectively) also underwent this tandem reaction as fast as **98a** (only 15 min at +50 °C in THF) and in good overall yields (formation of **105**, **106**, and **107**, respectively). When the migrating double bond is 1,2-disubstituted such as in **98e** (Scheme 37), the tandem sequence of isomerization–elimination still proceeds very efficiently and after transmetalation of the resulting dienyl zirconocene with copper salt, the allylation reaction gave the (*E,Z*)-triene **105** as unique isomer in 80% isolated yield. By combination of a long tether chain (six carbons) with a 1,2-disubstituted olefin as in **98f**, diene **108** was isolated in 61% yield. Furthermore, several different functionalizations of the resulting diene can also be performed such as the palladium-catalyzed cross-coupling reaction between dienylzirconocene **99Zr** and aryl iodide; **109** was obtained in moderate yield but as a unique isomer (Scheme 37). However, when the double bond is 1,1-disubstituted as in **98g**, or if an alkyl group is located in the carbon tether such as in **98h**, the reaction proceeds only in very low yield (Scheme 38).

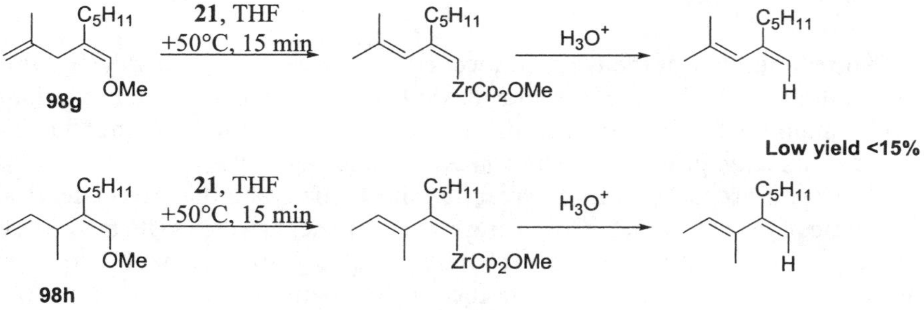

Scheme 38

The limitation of this methodology could be attributed not only to an unfavorable initial ligand exchange between (1-butene)ZrCp$_2$ **21** and the migrating olefin for steric reasons, but also to the formation of a hypothetically less stable trisubstituted zirconacyclopropane, **110** or **111**, which should be obtained after the first migration of the double bond (Scheme 39).

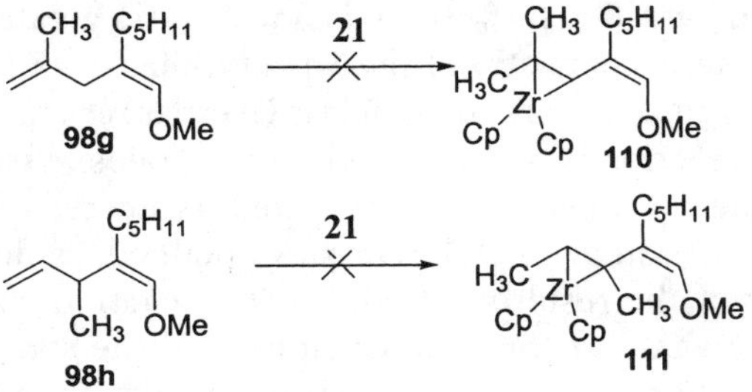

Scheme 39

In order to overcome these limitations, the more bulky bis(trimethylsilyl)-acetylene complex of zirconium (Rosenthal's complex) [75], which has recently offered a number of compelling advantages in synthesis [76], was tested in the reaction with **98 g** with the hope that the release of the bulky bis(trimethylsilyl)acetylene, after the ligand exchange, would be enough to drive the reaction to completion. However, under these conditions no isomerization–elimination was observed. Therefore, the precursor of a potentially more stable trisubstituted zirconacyclopropane **98i** was prepared and submitted to the isomerization–elimination sequence. After hydrolysis, three products **112**, *Z*-**113**, and *E*-**113** were obtained in a 2.3:3:1 ratio, respectively. Thus, when the migrating 1,1-disubstituted double bond is slightly activated toward the formation of the zirconacyclopropane (stabilization by a phenyl ring), the expected diene (besides the direct transformation of the methoxy enol ether into **112** after hydrolysis) could be formed but in this case, as a mixture of two geometrical isomers (Scheme 40).

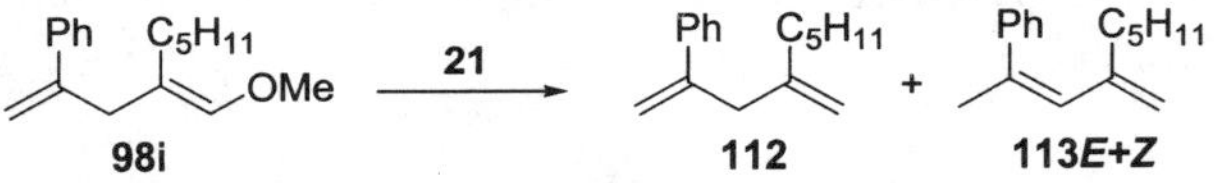

Scheme 40

A great number of dienyl zirconocenes are therefore formed by using this methodology, which presents a large flexibility (see Scheme 37), but limitations still remain for 1,1-disubstituted derivatives (as in **98g**) or compounds substituted by an alkyl group in the carbon tether (such as in **98h**).

To have more insight into the reaction mechanism and the stereochemical outcome of the reaction, the following two experiments were performed. First, it was checked that the reaction of trisubstituted enol ether with two alkyl groups, such as **114**, did not lead to the vinylic organometallic derivative [63] (Scheme 41), indicating that this tandem reaction should occur first by the isomerization of the remote double bond (only in the case of **98i**, the direct transformation of methoxy-enol ether into an organometallic derivative was

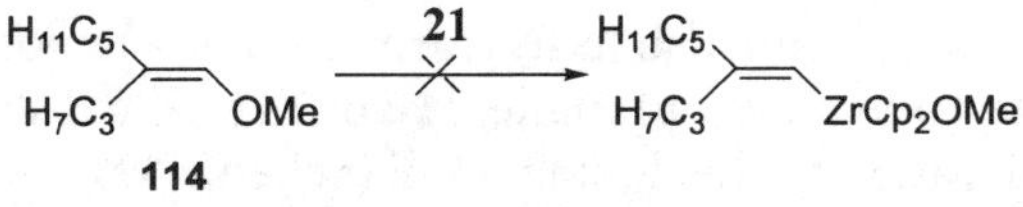

Scheme 41

Stereoselective Synthesis of Dienyl Zirconocene Complexes   157

detected, most probably due to a template effect between the zirconocene **21** and the nonmigrating double bond). Therefore, the mechanism of the isomerization was investigated with deuterium labeling experiments. When the dideutero enol ether **98j** (easily prepared according to Scheme 35) was treated with 1.3 equivalents of (1-butene)ZrCp$_2$ **21** in THF for 15 min at +50 °C, the corresponding 3,5-dideutero diene **115** was obtained in 62% isolated yield. The examination of the ^{1}H and ^{13}C NMR spectra indicates that, indeed, the two deuterium atoms are now located at the vinylic and allylic positions and led the authors to suggest the following mechanism for the allylic C–H bond activation–isomerization–elimination reaction (Scheme 42).

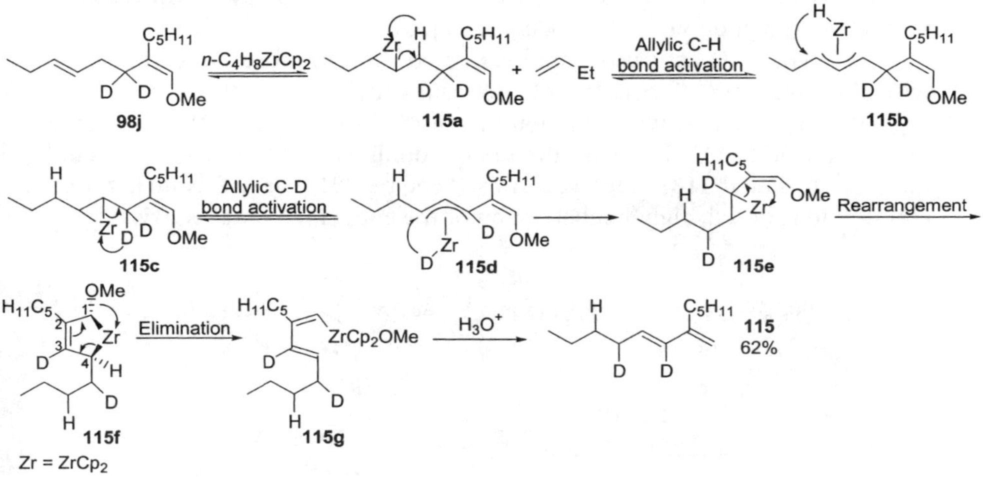

Scheme 42

(1-Butene)ZrCp$_2$ **21** reacts first with the remote double bond of **98j** to give the corresponding zirconacyclopropane **115a** and free butene. Then, via an allylic C–H bond activation [77], the η^3-allyl intermediate **115b** is generated as a transient species and, after hydrogen insertion, the new zirconacyclopropane **115c** is formed. By the same sequence, namely allylic C–D bond activation with deuterium migration (**115c** to η^3-allyl **115d** and then deuterium insertion), the zirconacyclopropane **115e** is produced. As soon as **115e** is formed, an irreversible step occurs transforming the zirconacyclopropane **115e** into zirconacyclopentene **115f**, which undergoes an elimination reaction to lead to **115g** and then **115** after hydrolysis. Based on this mechanism, the stereochemistry of the starting enol ether has no effect on the stereochemistry of the dienyl zirconocene; the carbon–heteroatom bond of the metalated center in **115f** can freely epimerize to give the most stable isomer. Such an isomerization could be caused by an interaction between the ether moiety and the zirconium atom, which would weaken the C$_1$–Zr bond and facilitate the isomerization [64]. In the particular experiment described in Scheme 42, no scrambling of deuterium atoms along the carbon skeleton was detected and this can be explained by the initial position of the two deuterium atoms. Indeed, in this allylic C–D bond activation step, as soon as the intermediate **115e** is formed, an irreversible rearrangement–elimination reaction (**115e** to **115g**) occurs, which therefore drives the reaction toward the metalated diene **115g**. However, when the two deuterium atoms are located in a different place in the tether, a scrambling of deuterium is observed along this tether.

The unique stereochemistry of the diene results therefore from the elimination step and not from a further isomerization of the dienyl zirconocene with zirconocene derivatives (i.e., Cp$_2$Zr-catalyzed stilbene stereoisomerization) [78]), since the hydrozirconation reaction of several 1*E*-ene-3-yne and 1*Z*-ene-3-yne derivatives with Cp$_2$ZrH(Cl) leads only to the (*E,E*) and (*Z,E*) isomers, respectively, in good yields (see Scheme 5) [33].

This allylic C–H activation can be, in some specific cases, in competition with the direct transformation of monosubstituted enol ether into vinylic organometallic derivatives. Compound **116** reacts faster with **21** by the enol moiety to lead to **117** than with the remote double bond of **116**, which would have given dienyl **118** after hydrolysis (Scheme 43). For the isomerization reaction to proceed, higher substitution of the enol ether is necessary.

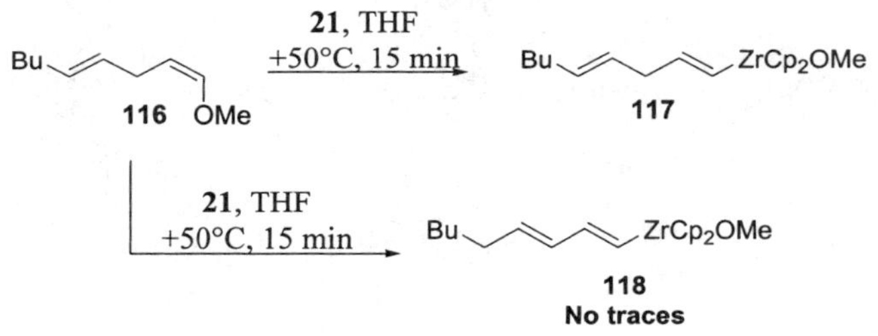

Scheme 43

In this tandem allylic C–H bond activation, followed by an elimination reaction, substituted 1-zircono-1*Z*,3*E*-dienes (zirconium moiety at the terminal position of the dienyl system) were easily prepared as unique isomers. With the idea of extending this methodology to the stereoselective synthesis of 3-zircono-1,3-diene (zirconium moiety at the internal position of the dienyl system), **119** was prepared and the reactivity was investigated with (1-butene)ZrCp$_2$ **21** (**119** was obtained by carbocupration of the α-allyl alkoxy-allene, Scheme 35) [79]. When **119** was submitted to the tandem reaction, the diene **120** was isolated after hydrolysis as a unique (*E,Z*) isomer in 75% isolated yield (Scheme 44).

When the migrating group is geminal to the leaving group such as in **119**, the allylic C–H bond activation leads to **120a**, which subsequently undergoes a β-elimination reaction to lead to the β-metalated allenyl intermediate **120b**. Then, **120b** is isomerized into its more stable dienyl form **120c** [80], in which the alkyl and the organometallic groups are *anti* to each other for steric reasons. After hydrolysis, a unique isomer is observed (determined by NOE). If *N*-bromosuccinimide is added as electrophile on **120c**, the corresponding bromodiene **121** is obtained as a single geometric isomer, as described in Scheme 45. The

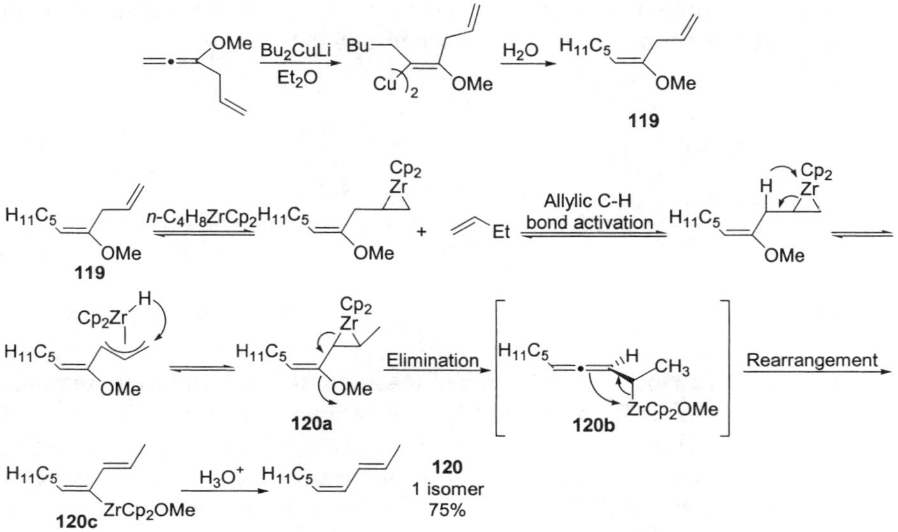

Scheme 44

stereochemistry of the reaction was also determined by addition of allyl chloride in the presence of a catalytic amount of copper salt (both CuCl/2LiCl and allyl chloride were added simultaneously and then heated at +60 °C for few hours), to give the dienyl zirconocene derivative **120c** [74]. Skipped triene **122** was obtained in good overall yield with an isomeric ratio greater than 98:2 in all cases. The (Z,E) stereochemistry of **122** was deduced on the basis of differential nuclear Overhauser effect spectra (Scheme 45).

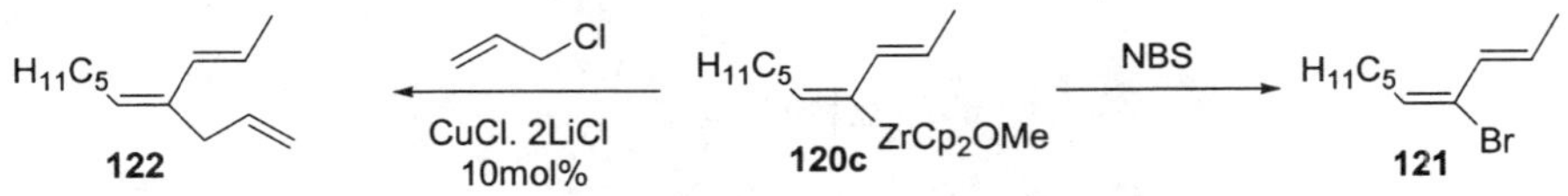

Scheme 45

Although this tandem reaction led also to the expected diene as a unique isomer in good chemical yield, this methodology had a serious drawback from a preparative point of view, since **119** (Scheme 44) could not be purified by column chromatography (purification of **119** invariably led to the corresponding ketone resulting from the hydrolysis of the enol ether moiety). The isolated yield obtained from **120** is therefore based on the crude starting material **119** used without purification. This very promising route for the preparation of metalated dienes such as **120c** associated with the problem of stability of α-substituted enol ether **119**, led the authors to consider an alternative starting material and therefore to the preparation of sulfonyl 1,3-dienyl derivatives [81]. As for the enol ether methodology, the main advantage of this approach is the very easy preparation of acyclic 2-arylsulfonyl 1,3-dienes **123a–d** from allylic sulfones and aldehydes in a single-pot operation, as described in Scheme 46 [82].

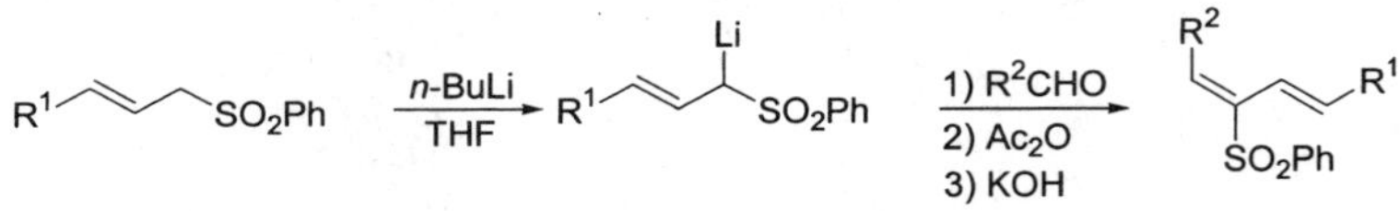

Scheme 46

123a R[1] = H, R[2] = Ph, 89%
123b R[1] = H, R[2] = n-Pent, 72%
123c R[1] = H, R[2] = c-Hex, 70%
123d R[1] = Me, R[2] = Ph, 91%

Treatment of **123b,c** with 1.5 equiv of (1-butene)ZrCp$_2$ **21** at room temperature leads only to the Z isomers **124** and **125**, whatever the stereochemistry of the starting dienyl sulfones (i.e., *E*-**123c** and *Z*-**123c**, Scheme 47). Even the unstable Z isomer **126** was preferentially formed from **123a** in this process with an excellent Z/E ratio of 95:5. The low yield obtained for **124** is attributed to the volatility of the resulting diene.

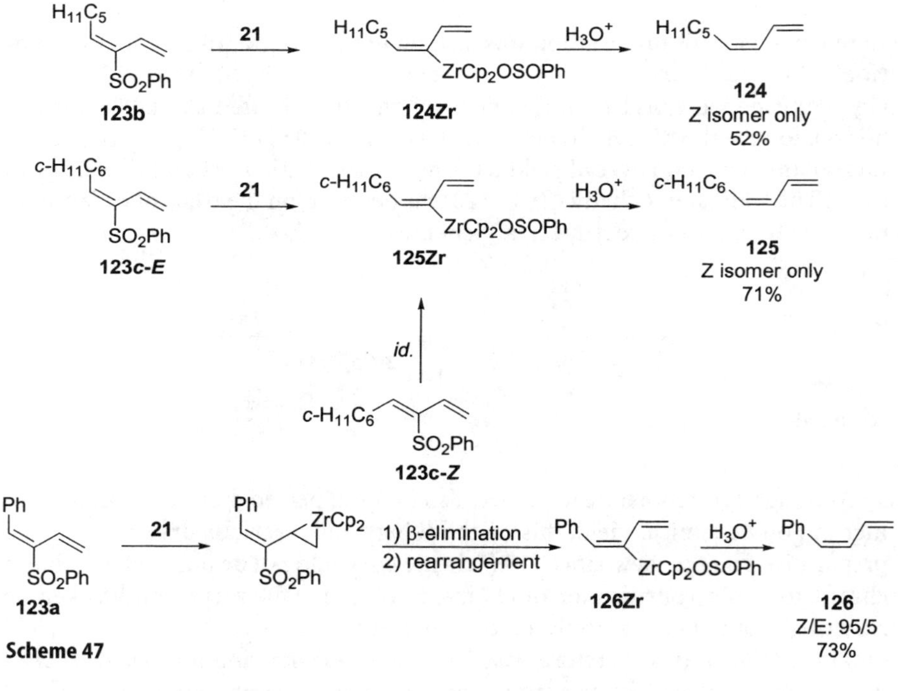

Scheme 47

By analogy with the mechanistic pathway described for the enol ether **119** (Scheme 44), we believe that the transformation of **123a–c** also occurs via the formation of the β-metalated allenyl intermediate, generated from the β-elimination of the corresponding zirconacyclopropane and subsequent rearrangement. However, the direct transformation of the vinyl sulfone moieties into dienyl zirconocenes without intervention of the terminal unsubstituted double bond cannot be ruled out at this stage [83]: when the more substituted dienyl sulfone **123d** (Scheme 48) was similarly treated with (1-butene)ZrCp$_2$ **21**, a mixture of three isomers of **127** of undetermined geometry was obtained after hydrolysis. Therefore, in this particular case, the method described in Scheme 44 gives better results.

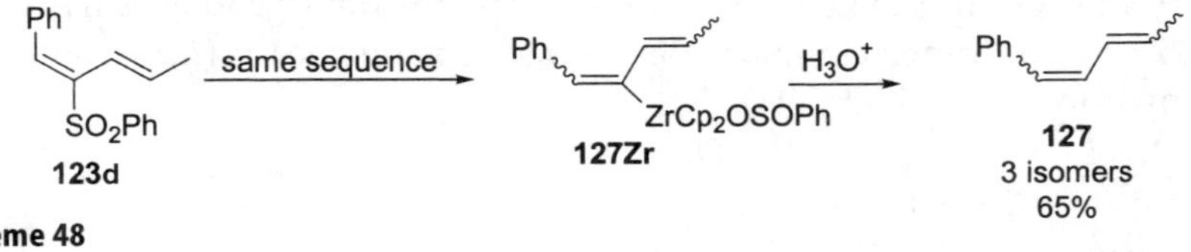

Scheme 48

As usual, to further increase the scope of the reaction, transmetalation of dienyl zirconium complexes, such as **124–127Zr**, into the corresponding dienyl organocopper derivatives was performed. Surprisingly, when **126Zr** was transmetalated to copper derivatives by addition of a catalytic amount of CuCl/2LiCl in the presence of allyl chloride for 1 h at +50 °C, a partial isomerization of the dienyl system was found (Scheme 49).

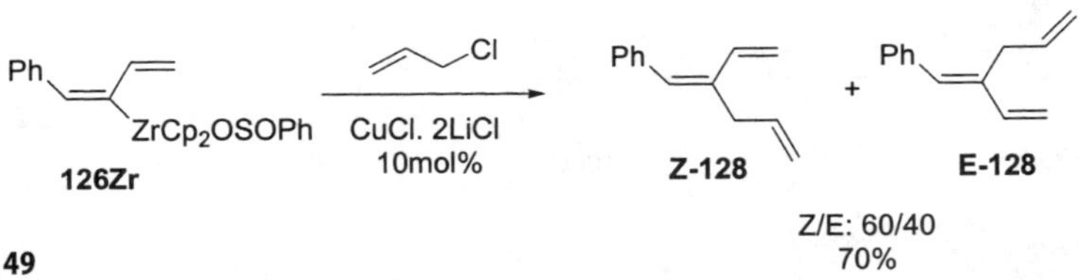

Scheme 49

This result is in contrast with those already obtained for the copper-catalyzed allylation reaction of **120c** into **122** (Scheme 45). This may be attributed to the very reactive nature of the diene **126Zr**. With the idea of having a complete isomerization reaction, **126Zr** was heated first at +50 °C in the presence of a stoichiometric amount of CuCl/2LiCl for 1 h before hydrolysis. Only the *E* isomer of **126** was isolated after this sequence meaning that *trans*-**126Zr** is transmetalated into *cis*-**126Cu** and then, after hydrolysis, only the *E* isomer of **126** is formed in 70% yield (Scheme 50). When heating the dienyl zirconocene **126Zr** at +50 °C, with or without added LiCl, no isomerization of the diene was detected.

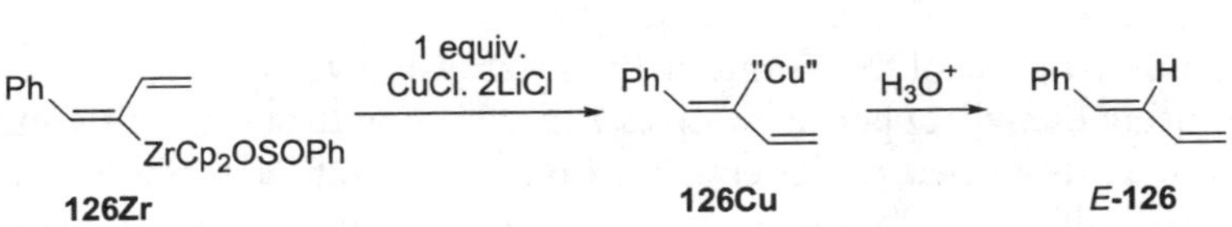

Scheme 50

As nothing is known about the exact nature of organocopper derivative **126Cu** coming from organozirconocene derivative **126Zr**, further investigations are needed to elucidate completely the mechanism of this transmetalation, but this isomerization was found to be general for all the examined cases (1 equiv of CuCl/2LiCl then +50 °C, 1 h; Scheme 51, **124Zr–127Zr**). It should be noted that even when a secondary alkyl group is present on the dienyl system, as in **125Zr**, the E isomer is the major isomer after the transmetalation step in a ratio of 86:14 (which implies that before hydrolysis, the copper is cis to the secondary alkyl group, Scheme 51). Moreover, from the three geometrical isomers of **127Zr**, this transmetalation–isomerization led mainly to the (E,E) isomer **127** after hydrolysis (E,E:Z,E=92:8).

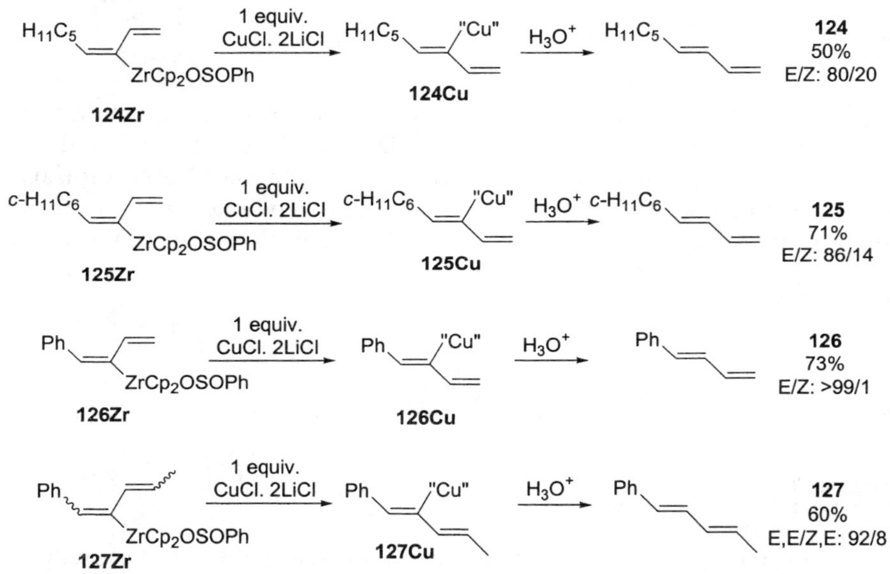

Scheme 51

Therefore, the isomerization reaction was also performed starting from the α-substituted enol ether **119** (Scheme 52). By using the tandem allylic C–H activation–elimination reactions, Z-**120c** is initially formed and by a transmetalation reaction into organocopper with a stoichiometric amount of CuCl/2LiCl, followed by heating at +50 °C for 1 h and reaction with allyl chloride, the resulting (E,E)-diene **122** is obtained with an isomeric ratio of 90:10 but in a low 40% yield as described in Scheme 52.

The synthetic use of this isomerization was also investigated by reaction of the resulting dienyl copper derivatives **124–126Cu** with several different electrophiles, as described in Scheme 53. Compound **126Cu** reacts via an S_N2' process with allyl chloride to give a unique E isomer of the skipped triene **128** and the geometrical mixture of **124Cu** and **125Cu** gave, under the same experimental

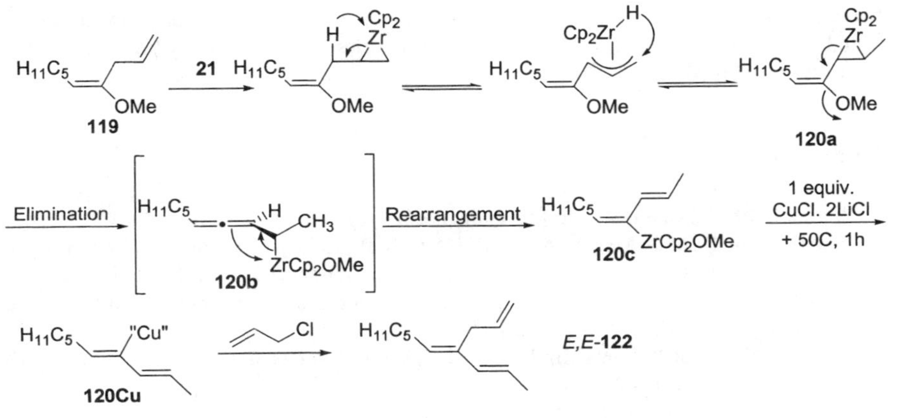

Stereoselective Synthesis of Dienyl Zirconocene Complexes

Scheme 52

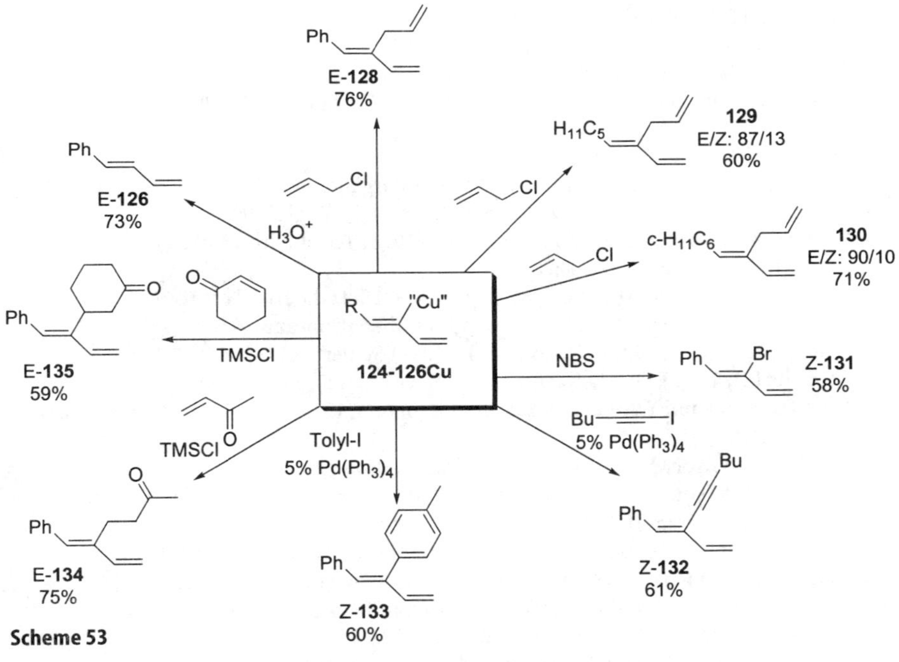

Scheme 53

conditions, the two allylated products **129** and **130** with the *E* isomer as major product. (It should be emphasized that although the transmetalation–isomerization occurs at +50 °C, the reactivity of the organocopper remains intact in this process). The addition of methyl vinyl ketone or cyclohexenone in the presence of TMSCl [84] led to the 1,4-adducts **134** and **135** in 75 and 59% yield, respectively, as unique geometrical isomers. The palladium cross-coupling reaction of **126Cu** with alkynyl iodide and aryl iodide opens new routes to further functionalization between two sp^2 and sp^2–sp units, as described for the preparation of **132** and **133**.

Acknowledgement This research was supported by the Israel Science Foundation administrated by the Israel Academy of Science and Humanities (79/01-1) and by the Fund for the Promotion of Research at the Technion.

References

1. Marek I (1999) J Chem Soc Perkin Trans I 535
2. Marek I (2004) In: de Meijere A, Diederich F (eds) Modern C–C and C–X bond formations by metal-catalyzed cross-coupling reactions. Wiley-VCH, Weinheim (in press)
3. Marek I (2004) In: Beller M, Bolm C (eds) Transition metals for organic synthesis, 2nd edn. Wiley-VCH, Weinheim (in press)
4. Marek I, Normant JF (1998) In: Diederich F, Stang PJ (eds) Metal-catalyzed cross-coupling reactions. Wiley-VCH, Weinheim, p 271
5. Normant JF, Alexakis A (1981) Synthesis 841
6. Knochel P (1991) In: Trost BM, Fleming I, Semmelhack MF (eds) Comprehensive organic synthesis. Pergamon, New York, p 865
7. Negishi E (1981) Pure Appl Chem 53:2333
8. Fallis AG, Forgione P (2001) Tetrahedron 57:5899
9. Negishi E, Takahashi T (1988) Synthesis 1
10. Negishi E, Montchamp JL, Anastasia L, Alizarov A, Choueiry D (1998) Tetrahedron Lett 39:2503
11. Ma S, Negishi E (1997) J Org Chem 62:784
12. Negishi E, Kondakov DY, Van Horn DE (1997) Organometallics 16:951
13. Studemann T, Knochel P (1997) Angew Chem Int Ed Engl 36:93
14. Studemann T, Ibrahim-Ouali M, Knochel P (1998) Tetrahedron 54:1299
15. Marek I (2000) Chem Rev 100:2887
16. Yamanoi S, Imai T, Masumoto T, Suzuki K (1997) Tetrahedron Lett 38:3031
17. Yamaguchi M, Sotokawa T, Hirama M (1997) Chem Commun 743
18. Usugi S, Tang J, Shinokubo H, Oshima K (1999) Synlett 1417
19. Yorimitsu H, Tang J, Okada K, Shinokubo H, Oshima K (1998) Chem Lett 11
20. Hojo M, Murakami Y, Aihara H, Sakuragi R, Baba Y, Hosomi A (2001) Angew Chem Int Ed 40:621
21. Forgione P, Wilson PD, Fallis AG (2000) Tetrahedron Lett 41:17
22. Forgione P, Wilson PD, Yap GPA, Fallis AG (2000) Synthesis 921
23. Alexakis A, Normant JF (1982) Tetrahedron Lett 23:5151
24. Furber M, Taylor RJK, Burford SC (1986) J Chem Soc Perkin Trans I 1809
25. Takahashi T, Kondakov DY, Xi Z, Suzuki NA (1995) J Am Chem Soc 117:5871
26. Hyuga S, Yamashina N, Hara S, Suzuki A (1988) Chem Lett 29:1811
27. Negishi E (1982) Acc Chem Res 15:340
28. Sharma S, Oehlschlager AC (1988) Tetrahedron Lett 29:261
29. Hata T, Kitagawa H, Masai H, Kurahashi T, Shimizu M, Hiyama T (2001) Angew Chem Int Ed 40:790
30. Buchwald SL, LaMaire SJ, Nielsen RB, Watson BT, King SM (1987) Tetrahedron Lett 28:3895
31. Lipshutz BH, Steven SP, Noson K, Tomioka T (2002) In: Marek I (ed) Titanium and zirconium in organic synthesis. Wiley-VCH, Weinheim, p 110
32. Fryzuk MD, Bates GS, Stone C (1991) J Org Chem 56:7201
33. Fryzuk MD, Bates GS, Stone C (1986) Tetrahedron Lett 27:1537
34. Buchwald SL, Waston BT, Huffman JC (1987) J Am Chem Soc 109:2544

35. Buchwald SL, Nielsen RB (1989) J Am Chem Soc 111:2870
36. Negishi E, Takahashi T (1998) Bull Chem Soc Jpn 71:755
37. Negishi E, Takahashi T (1994) Acc Chem Res 27:124
38. Negishi E, Kondakov DY (1996) Chem Soc Rev 26:417
39. Nielsen RB, Buchwald SL (1988) Chem Rev 88:1047
40. Takahashi T, Li Y (2002) In: Marek I (ed) Titanium and zirconium in organic synthesis. Wiley-VCH, Weinheim, p 50
41. Nugent WA, Thorn DL, Harlow RL (1987) J Am Chem Soc 109:2788
42. Fagan PJ, Nugent WA (1988) J Am Chem Soc 110:2312
43. Negishi E, Holmes SJ, Tour JM, Miller JA, Cederbaum FE, Swanson DR, Takahashi T (1989) J Am Chem Soc 111:3336
44. Hara R, Nishihara Y, Landre PD, Takahashi T (1997) Tetrahedron Lett 38:447
45. Ubayama H, Xi Z, Takahashi T (1998) Chem Lett 517
46. Leng L, Xi C, Chen C, Lai C (2004) Tetrahedron Lett 45:595
47. Suzuki N, Konadkov DY, Kageyama M, Kotora M, Hara R, Takahashi T (1995) Tetrahedron 51:4519
48. Takahashi T, Hara R, Huo S, Ura Y, Leese MP, Suzuki N (1997) Tetrahedron Lett 38:8723
49. Takahashi T, Kondakov DY, Xi Z, Suzuki N (1995) J Am Chem Soc 117:5871
50. Barluenga J, Rodriguez F, Alvarez-Rodrigo L, Fañanas FJ (2004) Chem Eur J 10:101
51. Kociensky P, Barber C (1990) Pure Appl Chem 62:1993
52. Shimizu N, Shibata F, Tsuno Y (1987) Bull Chem Soc Jpn 60:777
53. Kasatkin A, Whitby RJ (1999) J Am Chem Soc 121:7039
54. Kasatkin A, Whitby RJ (1997) Tetrahedron Lett 38:4857
55. Negishi E, Okukado N, King AO, Van Horn DE, Spiegel BI (1978) J Am Chem Soc 100:2254
56. Matsushita H, Negishi E (1981) J Am Chem Soc 103:2882
57. Negishi E, Takahashi T, Baba S, Van Horn DE, Okukado N (1987) J Am Chem Soc 109:2393
58. Topolsky M, Duraisamy M, Rachon J, Gawronskyi J, Gawronska K, Goedken V, Walborsky HM (1993) J Org Chem 58:546
59. Nelson DJ, Matthews MKG (1994) J Organomet Chem 469:1
60. Schleyer PvR, Clark T, Kos AJ, Spitznagel GW, Rohde C, Arad D, Houk KN, Rondan NG (1984) J Am Chem Soc 106:6467
61. Ganchegui B, Bertus P, Szymoniak J (2001) Synlett 123
62. Negishi E, Choueiry D, Nguyen TB, Swanson DR, Suzuki N, Takahashi T (1994) J Am Chem Soc 116:9751
63. Liard A, Marek I (2000) J Org Chem 65:7218
64. Takahashi T, Fujimoto T, Seki T, Saburi M, Uchida Y, Rousset CJ, Negishi E, (1990) J Chem Soc Chem Commun 182
65. (a) Mintz EA, Ward AS, Tice DS (1985) Organometallics 4:1308; (b) Ward AS, Mintz EA, Kramer MP (1988) Organometallics 7:8
66. (a) Takahashi T, Suzuki N, Kageyama M, Nitto Y, Saburi M, Negishi E (1991) Chem Lett 1579; (b) Takahashi T, Nitto Y, Seki T, Saburi M, Negishi E (1990) Chem Lett 2259
67. (a) Tolman CA (1972) J Am Chem Soc 94:2994; (b) Bingham D, Hudson B, Webster BDE, Wells PB (1974) J Chem Soc Dalton Trans 1521; (c) Bingham D, Webster BDE, Wells PB (1974) J Chem Soc Dalton Trans 1514
68. Akita M, Yasuda H, Nagasuna K, Nakamura A (1983) Bull Chem Soc Jpn 56:554
69. (a) Swanson DR, Negishi E (1991) Organometallics 10:825; (b) Maye JP, Negishi E (1993) Tetrahedron Lett 34:3359; (c) Negishi E, Maye JP, Choueiry D (1995) Tetrahedron 51:4447
70. Brandsma L (1981) Synthesis of acetylenes allenes and cumulenes. Elsevier, Amsterdam
71. Alexakis A, Normant JF (1984) Isr J Chem 24:113

72. Takahashi T, Fujimori T, Seki T, Saburi M, Uchida Y, Rousset CJ, Negishi E (1990) J Chem Soc Chem Commun 182
73. (a) Chinkov N, Majumdar S, Marek I (2002) J Am Chem Soc 124:10282; (b) Chinkov N, Majumdar S, Marek I (2003) J Am Chem Soc 125:13258
74. (a) Wipf P, Jahn H (1996) Tetrahedron 52:12853; (b) Takahashi T, Kotora M, Kasai K, Suzuki N (1994) Tetrahedron Lett 35:5685
75. Rosenthal U, Burlakov VV (2002) In: Marek I (ed) Titanium and zirconium in organic synthesis. Wiley-VCH, Weinheim, p 355
76. Nitschke JR, Zurcher S, Tilley DT (2000) J Am Chem Soc 122:10345
77. (a) Resconi L (1999) J Mol Catal 146:167; (b) Cohen SA, Auburn PR, Bercaw JE (1983) J Am Chem Soc 105:1136
78. Takahashi T, Swanson DR, Negishi E (1987) Chem Lett 623
79. Clinet JC, Linstrumelle G (1978) Tetrahedron Lett 1137
80. Rozema MJ, Knochel P (1991) Tetrahedron Lett 32:1855
81. Backvall JE, Chinchilla R, Najera C, Yus M (1998) Chem Rev 98:2291
82. Cuvigny T, Herve du Penhoat C, Julia M (1986) Tetrahedron 42:5329
83. (a) Farhat S, Marek I (2002) Angew Chem Int Ed 41:1410; (b) Faraht S, Zouev I, Marek I (2004) Tetrahedron 60:1329; (c) Chinkov N, Chechik H, Majumdar S, Liard A, Marek I (2002) Synthesis 2473
84. (a) Nakamura E, Kuwajima I (1984) J Am Chem Soc 106:3368; (b) Corey EJ, Boaz NW (1985) Tetrahedron Lett 26:6015; (c) Alexakis A, Berlan J, Besace Y (1986) Tetrahedron Lett 27:2143

Author Index Volumes 1-10

Volume 9 is already in planning and is announced and will be published when the manuscripts are submitted to the publisher.
The volume numbers are printed in italics.

Abdel-Magid AF see Mehrmann SJ (2004) 6: 153-180
Alper H see Grushin VV (1999) 3: 193-225
Anwander R (1999) Principles in Organolanthanide Chemistry. 2: 1-62
Armentrout PB (1999) Gas-Phase Organometallic Chemistry 4: 1-45

Beak P, Johnson TA, Kim DD, Lim SH (2003) Enantioselective Synthesis by Lithiation Adjacent to Nitrogen and Electrophile Incorporation. 5: 139-176
Bertus P see Szymoniak J (2005) 10: 107-132
Bien J, Lane GC, Oberholzer MR (2004) Removal of Metals from Process Streams: Methodologies and Applications. 6: 263-284
Böttcher A see Schmalz HG (2004) 7: 157-180
Braga D (1999) Static and Dynamic Structures of Organometallic Molecules and Crystals. 4: 47-68
Brüggemann M see Hoppe D (2003) 5: 61-138

Chlenov A see Semmelhack MF (2004) 7: 21-42
Chlenov A see Semmelhack MF (2004) 7: 43-70
Chinkov M, Marek I (2005) Stereoselective Synthesis of Dienyl Zirconocene Complexes. 10: 133-166
Clayden J (2003) Enantioselective Synthesis by Lithiation to Generate Planar or Axial Chirality. 5: 251-286
Cummings SA, Tunge JA, Norton JR (2005) Synthesis and Reactivity of Zirconaaziridines. 10: 1-39

Dedieu A (1999) Theoretical Treatment of Organometallic Reaction Mechanisms and Catalysis. 4: 69-107
Delmonte AJ, Dowdy ED, Watson DJ (2004) Development of Transition Metal-Mediated Cyclopropanation Reaction. 6: 97-122
Dowdy EC see Molander G (1999) 2: 119-154
Dowdy ED see Delmonte AJ (2004) 6: 97-122

Eisen MS see Lisovskii A (2005) 10: 63-105

Fürstner A (1998) Ruthenium-Catalyzed Metathesis Reactions in Organic Synthesis. 1: 37-72

Gibson SE (née Thomas), Keen SP (1998) Cross-Metathesis. 1: 155-181
Gisdakis P see Rösch N (1999) 4: 109-163

Görling A see Rösch N (1999) 4: 109–163
Goldfuss B (2003) Enantioselective Addition of Organolithiums to C=O Groups and Ethers. 5: 12–36
Gossage RA, van Koten G (1999) A General Survey and Recent Advances in the Activation of Unreactive Bonds by Metal Complexes. 3: 1–8
Gotov B see Schmalz HG (2004) 7: 157–180
Gras E see Hodgson DM (2003) 5: 217–250
Grepioni F see Braga D (1999) 4: 47–68
Gröger H see Shibasaki M (1999) 2: 199–232
Grushin VV, Alper H (1999) Activation of Otherwise Unreactive C–Cl Bonds. 3: 193–225

Harman D (2004 Dearomatization of Arenes by Dihapto-Coordination. 7: 95–128
He Y see Nicolaou KC, King NP (1998) 1: 73–104
Hidai M, Mizobe Y (1999) Activation of the N–N Triple Bond in Molecular Nitrogen: Toward its Chemical Transformation into Organo-Nitrogen Compounds. 3: 227–241
Hodgson DM, Stent MAH (2003) Overview of Organolithium-Ligand Combinations and Lithium Amides for Enantioselective Processes. 5: 1–20
Hodgson DM, Tomooka K, Gras E (2003) Enantioselective Synthesis by Lithiation Adjacent to Oxygen and Subsequent Rearrangement. 5: 217–250
Hoppe D, Marr F, Brüggemann M (2003) Enantioselective Synthesis by Lithiation Adjacent to Oxygen and Electrophile Incorporation. 5: 61–138
Hou Z, Wakatsuki Y (1999) Reactions of Ketones with Low-Valent Lanthanides: Isolation and Reactivity of Lanthanide Ketyl and Ketone Dianion Complexes. 2: 233–253
Hoveyda AH (1998) Catalytic Ring-Closing Metathesis and the Development of Enantioselective Processes. 1: 105–132
Huang M see Wu GG (2004) 6: 1–36
Hughes DL (2004) Applications of Organotitanium Reagents. 6: 37–62

Iguchi M, Yamada K, Tomioka K (2003) Enantioselective Conjugate Addition and 1,2-Addition to C=N of Organolithium Reagents. 5: 37–60
Ito Y see Murakami M (1999) 3: 97–130
Ito Y see Suginome M (1999) 3: 131–159

Jacobsen EN see Larrow JF (2004) 6: 123–152
Johnson TA see Break P (2003) 5: 139–176
Jones WD (1999) Activation of C–H Bonds: Stoichiometric Reactions. 3: 9–46

Kagan H, Namy JL (1999) Influence of Solvents or Additives on the Organic Chemistry Mediated by Diiodosamarium. 2: 155–198
Kakiuchi F, Murai S (1999) Activation of C–H Bonds: Catalytic Reactions. 3: 47–79
Kanno K see Takahashi T (2005) 8: 217–236
Keen SP see Gibson SE (née Thomas) (1998) 1: 155–181
Kendall C see Wipf P (2005) 8: 1–25
Kiessling LL, Strong LE (1998) Bioactive Polymers. 1: 199–231
Kim DD see Beak P (2003) 5: 139–176
King AO, Yasuda N (2004) Palladium-Catalyzed Cross-Coupling Reactions in the Synthesis of Pharmaceuticals. 6: 205–246
King NP see Nicolaou KC, He Y (1998) 1: 73–104
Kobayashi S (1999) Lanthanide Triflate-Catalyzed Carbon–Carbon Bond-Forming Reactions in Organic Synthesis. 2: 63–118

Kobayashi S (1999) Polymer-Supported Rare Earth Catalysts Used in Organic Synthesis. 2: 285–305
Kodama T see Arends IWCE (2004) 11: 277–320
Kondratenkov M see Rigby J (2004) 7: 181–204
Koten G van see Gossage RA (1999) 3: 1–8
Kotora M (2005) Metallocene-Catalyzed Selective Reactions. 8: 57–137
Kumobayashi H, see Sumi K (2004) 6: 63–96
Kündig EP (2004) Introduction 7: 1–2
Kündig EP (2004) Synthesis of Transition Metal η^6-Arene Complexes. 7: 3–20
Kündig EP, Pape A (2004) Dearomatization via η^6 Complexes. 7: 71–94

Lane GC see Bien J (2004) 6: 263–284
Larrow JF, Jacobsen EN (2004) Asymmetric Processes Catalyzed by Chiral (Salen)Metal Complexes 6: 123–152
Li Z, see Xi Z (2005) 8: 27–56
Lim SH see Beak P (2003) 5: 139–176
Lin Y-S, Yamamoto A (1999) Activation of C–O Bonds: Stoichiometric and Catalytic Reactions. 3: 161–192
Lisovskii A, Eisen MS (2005) Octahedral Zirconium Complexes as Polymerization Catalysts. 10: 63–105

Marek I see Chinkov M (2005) 10: 133–166
Marr F see Hoppe D (2003) 5: 61–138
Maryanoff CA see Mehrmann SJ (2004) 6: 153–180
Maseras F (1999) Hybrid Quantum Mechanics/Molecular Mechanics Methods in Transition Metal Chemistry. 4: 165–191
Medaer BP see Mehrmann SJ (2004) 6: 153–180
Mehrmann SJ, Abdel-Magid AF, Maryanoff CA, Medaer BP (2004) Non-Salen Metal-Catalyzed Asymmetric Dihydroxylation and Asymmetric Aminohydroxylation of Alkenes. Practical Applications and Recent Advances. 6: 153–180
Mizobe Y see Hidai M (1999) 3: 227–241
Molander G, Dowdy EC (1999) Lanthanide- and Group 3 Metallocene Catalysis in Small Molecule Synthesis. 2: 119–154
Mori M (1998) Enyne Metathesis. 1: 133–154
Mori M (2005) Synthesis and Reactivity of Zirconium-Silene Complexes. 10: 41–62
Muñiz K (2004) Planar Chiral Arene Chromium (0) Complexes as Ligands for Asymmetric Catalysis. 7: 205–223
Murai S see Kakiuchi F (1999) 3: 47–79
Murakami M, Ito Y (1999) Cleavage of Carbon–Carbon Single Bonds by Transition Metals. 3: 97–130

Nakamura S see Toru T (2003) 5: 177–216
Namy JL see Kagan H (1999) 2: 155–198
Negishi E, Tan Z (2005) Diastereoselective, Enantioselective, and Regioselective Carboalumination Reactions Catalyzed by Zirconocene Derivatives. 8: 139–176
Nicolaou KC, King NP, He Y (1998) Ring-Closing Metathesis in the Synthesis of Epothilones and Polyether Natural Products. 1: 73–104
Normant JF (2003) Enantioselective Carbolithiations. 5: 287–310
Norton JR see Cummings SA (2005) 10: 1–39

Oberholzer MR see Bien J (2004) 6: 263–284

Pape A see Kündig EP (2004) 7: 71–94
Pawlow JH see Tindall D, Wagener KB (1998) 1: 183–198
Prashad M (2004) Palladium-Catalyzed Heck Arylations in the Synthesis of Active Pharmaceutical Ingredients. 6: 181–204

Richmond TG (1999) Metal Reagents for Activation and Functionalization of Carbon–Fluorine Bonds. 3: 243–269
Rigby J, Kondratenkov M (2004) Arene Complexes as Catalysts. 7: 181–204
Rodríguez F see Barluenga (2004) 13: 59–121
Rösch N (1999) A Critical Assessment of Density Functional Theory with Regard to Applications in Organometallic Chemistry. 4: 109–163
Schmalz HG, Gotov B, Böttcher A (2004) Natural Product Synthesis. 7: 157–180
Schrock RR (1998) Olefin Metathesis by Well-Defined Complexes of Molybdenum and Tungsten. 1: 1–36
Semmelhack MF, Chlenov A (2004) (Arene)Cr(Co)$_3$ Complexes: Arene Lithiation/Reaction with Electrophiles. 7: 21–42
Semmelhack MF, Chlenov A (2004) (Arene)Cr(Co)$_3$ Complexes: Aromatic Nucleophilic Substitution. 7: 43–70
Sen A (1999) Catalytic Activation of Methane and Ethane by Metal Compounds. 3: 81–95
Shibasaki M, Gröger H (1999) Chiral Heterobimetallic Lanthanoid Complexes: Highly Efficient Multifunctional Catalysts for the Asymmetric Formation of C–C, C–O and C–P Bonds. 2: 199–232
Stent MAH see Hodgson DM (2003) 5: 1–20
Strong LE see Kiessling LL (1998) 1: 199–231
Suginome M, Ito Y (1999) Activation of Si–Si Bonds by Transition-Metal Complexes. 3: 131–159
Sumi K, Kumobayashi H (2004) Rhodium/Ruthenium Applications. 6: 63–96
Suzuki N (2005) Stereospecific Olefin Polymerization Catalyzed by Metallocene Complexes. 8: 177–215
Szymoniak J, Bertus P (2005) Zirconocene Complexes as New Reagents for the Synthesis of Cyclopropanes. 10: 107–132

Takahashi T, Kanno K (2005) Carbon-Carbon Bond Cleavage Reaction Using Metallocenes. 8: 217–236
Tan Z see Negishi E (2005) 8: 139–176
Tindall D, Pawlow JH, Wagener KB (1998) Recent Advances in ADMET Chemistry. 1: 183–198
Tomioka K see Iguchi M (2003) 5: 37–60
Tomooka K see Hodgson DM (2003) 5: 217–250
Toru T, Nakamura S (2003) Enantioselective Synthesis by Lithiation Adjacent to Sulfur, Selenium or Phosphorus, or without an Adjacent Activating Heteroatom. 5: 177–216
Tunge JA see Cummings SA (2005) 10: 1–39

Uemura M (2004) (Arene)Cr(Co)$_3$ Complexes: Cyclization, Cycloaddition and Cross Coupling Reactions. 7: 129–156

Wagener KB see Tindall D, Pawlow JH (1998) 1: 183–198
Wakatsuki Y see Hou Z (1999) 2: 233–253
Watson DJ see Delmonte AJ (2004) 6: 97–122
Wipf P, Kendall C (2005) Hydrozirconation and Its Applications. 8: 1–25
Wu GG, Huang M (2004) Organolithium in Asymmetric Process. 6: 1–36

Xi Z, Li Z (2005) Construction of Carbocycles via Zirconacycles and Titanacycles. 8: 27–56

Yamada K see Iguchi M (2003) 5: 37–60
Yamamoto A see Lin Y-S (1999) 3: 161–192
Yasuda H (1999) Organo Rare Earth Metal Catalysis for the Living Polymerizations of Polar and Nonpolar Monomers. 2: 255–283
Yasuda N see King AO (2004) 6: 205–246

Subject Index

Acetylenes 15
Active sites, uniformity 91
Alcohol, homoallylic 109
Aldehydes, imines, zirconaziridines 21
Alkenyl carbenoids 145
Alkoxy-allene 152
Allenyl intermediate, β-metalated 158
Allylation, copper-catalyzed 161
Allylcyclopropane 115
Allylic activation 133
Allylmetalation 134
Allylsilene 50
Aluminacycle 126, 127
Aluminacyclopentane 127
Aluminacyclopentene 127
Atactic domains 92
Atactic polymer 66
Atomic force microscopy (AFM) 100
Azaallyl hydride 31–36
Azazirconacyclopropane 47

Benzamidinate complexes 63

^{13}C-NMR triad/pentad analysis 69
C_2-symmetry complexes 65
C_{2v}-symmetry 92
C-D bond activation, allylic 157
C-H bond activation, allylic 157
Carbenoid 109
Carboalumination 127, 128
Carbocyclization, intramolecular 140
Carbodiimides 22
Carbometalation 134
Carbometalative ring expansion 149
Carbonylation 58
Carbon-zirconium bond 141
Catalysts, heterogenization 80
–, supported, homogeneous 85
Catalytic precursor 65

Chlorosilylzirconocene 51, 55
Circular dichroism 34
Cocatalysts 64, 66
Coupling, oxidative 91
Cross-coupling reactions 135
Crystalline domains 100
Curtin-Hammett 2, 26, 30
Cyclobutane 124, 125
Cyclopropanol 108, 110
Cyclopropylamine 108
Cyclopropylcarbinyl-homoallyl rearrangement 117
Cyclopropylcarbinylzirconium complex 117

Decarbozirconation 150
Deoxygenative process 108, 109, 129
Deuterium insertion/migration 157
Deuterium labeling experiments 157
Dibutylzirconocene 52
Dienes 135
–, 2-zirconated 146
Dienyl copper derivatives 162
Dienyl zirconium 136
Dienyl zirconocene 133, 153
Diffusion effect 86
1,4-Dihalogenodiene 141
Disilylzirconocene 52, 55
Diynes, bicyclization 140
Dynamic kinetic asymmetric transformation 2, 30, 31, 36
Dynamic resolution, kinetic 29
– –, thermodynamic 27

EBTHI zirconaziridines 17
Elastomers 66
Elimination 133, 142, 155
–, beta 114, 119
–, β-methyl 75

Subject Index

Elimination
–, gamma 114–117, 122
Enol ether 155
Enynes, hydrozirconation 133
Epimerization 70
Ethenylzirconation, vinyl ethers 142
Ethylene 138

Functionalizations 155

1-Halo-1-lithioalkene 145
Heteroallylic compounds 65
Heterocumulenes 14, 22
HMS silica 81
Homoenolate anion equivalent 114, 123
Hydrozirconation 117–119, 128, 137

Iminosilacyl zirconium complex 46, 47
Iminosilazirconacyclohexene 56–58
Imino-zirconium complex 56, 57
Inductive effect 69
Insertion, isonitrile 56
–, stereospecific 146
Inversion 145
Ionization, metal-assisted 147
Iridium-silene complex 44
Iron-silene complex 43
Isocyanates 22
Isocyanide 7
Isomerization 133, 150, 151, 153
Isotactic domains 85
Isotactic polymer 66

Lewis acid 112, 120, 124

MAO (methylalumoxane) 64
MCM-41 81
1,2-Metalate rearrangement 146
Metallacyclopentadiene 137
Metallocenes 64
–, supported 80
Methyl silyl ketone 58
Methylalumoxane 64
Molecular weight distribution 91

Negishi reagent 6, 138

Olefin coordination, Dewar-Chatt-Duncanson 10

Olefins, heterosubstituted 148
–, α-, isomerization 73
Organozirconocene derivatives, carbenoid reagents 133
Oxazirconacyclohexene 58

Palladium cross-coupling reaction 163
Phosphinoamide moieties 63
Polymerization, stereoregular 64
Polypropylene, elastomeric 87
–, –, selective design 99
–, stereodefects 73
Propeller-like structure 76
Propylene 63
Proton/carbon spectroscopic analysis 96

Racemic mixtures 65
Ring contraction 110
Rosenthal reagent 6
Rosenthal's complex 156
Ruthenium-silene complex 44

Schwartz reagent 136
Silacylzirconium complex 46
Silazirconacyclohexenone 58
Silazirconacyclopentene 47, 52, 55–58
Silazirconacyclopropane 52, 58
Silene 42
Silene-silylene rearrangement 45
Silica 80
–, mesoporous 81
Silicon-carbon bond 42, 45
Silicon-zirconium bond 45
Silyl lithium 42, 47, 48
Silylmethylaniline 47
2-Silyloxy-1,3-dienes 147
Silyl-zirconium complex 45
Skeletal rearrangement 150, 153
Solids, mesoporous 80
Stereoerrors 85
Stereoisomerization 149
Stereospecific polymerization 64
Sulfonyl 1,3-dienyl derivatives 159

Tandem sequence 155
Termination chain mechanism 72
Tetrahedral-octahedral equilibrium 97
Titanocene, low-valent 151
Transition metal-silene complex 43
Transmetalation 59, 125, 128, 162

Subject Index

Transmetalation-isomerization 162
Triene, skipped 153
Trimethylsilyl 65
Tungsten-silene complex 43

Umpolung 2

Vinyl sulfone 161
E-Vinyl zirconium 149
Vinylcyclopropane 110, 111
Vinylmetalation 134
Vinylsilane 48

Zeolites 80
Ziegler-Natta 64
Zirconaaziridines 1
Zirconacycle 55
–, substituted 141

Zirconacyclopentadiene 126, 133, 137
Zirconacyclopentane 115, 138
Zirconacyclopentene 55
Zirconacyclopropane 55, 110–112, 149
Zirconacyclopropene 136
Zirconium, oxidation states 90
Zirconium benzamidinate 80
Zirconium-carbon bond 45, 52, 56
Zirconium complexes, octahedral 63
Zirconium hydride 136
Zirconium-silene complex 52, 55, 58
Zirconocene 52, 55
Zirconocene chloride 145
Zirconocene-ethene complex 139
Zirconocene methyl chloride 3, 143
3-Zircono-1,3-diene 158

Printing: Krips bv, Meppel
Binding: Litges & Dopf, Heppenheim